THE SCHOOL POISONING TRAGEDY IN CALEDONIA, OHIO

DR. JAMES VAN KEUREN

THE History PRESS

Published by The History Press
Charleston, SC
www.historypress.com

Cover images courtesy of the *Columbus Dispatch*.

First published 2021

Manufactured in the United States
ISBN 9781467146326

Library of Congress Control Number: 2020945746

To the graduates of the River Valley Local School District in Caledonia, Ohio, who battled cancer and were left with a lifetime of pain and suffering, and to those who passed away at such a young age. I believe that the information and the data points presented in the book raise questions about the toxic contamination found throughout the school grounds. This is not a report clouded by scientific and technical jargon, but a presentation of understandable facts that gives the reader the opportunity to judge the validity of the conclusions reached by the government agencies that found no link between the contaminants on the school grounds to the various incidents of cancers among the River Valley High School graduates.

CONTENTS

PREFACE

This story begins with the release of my recent book *World War II POW Camps in Ohio*, which describes in a chapter called "Contamination on the Home Front" the various former military sites used during World War II in and around Camp Perry, Ohio, especially the Marion Engineer Depot, Marion, Ohio, which were found to be contaminated after their abandonment by the military. One such site was the River Valley Local School District in Caledonia, Ohio. Researching that book triggered my curiosity about the prevalence of contaminated sites across the United States. I discovered from the U.S. Army Corps of Engineers site map that there are 5,144 Formerly Used Defense Sites, including 95 in my home state of Ohio.[1] Across the United States and its territories, some 1,337 Superfund sites have known releases or threat of hazardous substances, pollutants or contaminants, and an additional 68 partial completions and 1,206 construction completions bring the total to 2,543 Superfund sites, 38 of them in Ohio.[2]

I started to speculate whether the communities surrounding the contaminated sites were exposed to toxic waste and whether such exposure contributed to debilitating diseases. In fact, the nationally known contamination sites with high cancer rates include Woburn, Massachusetts, and Toms River, New Jersey, where industrial waste caused contaminated drinking water supplies. Then there is Camp Lejeune, where the drinking water had been contaminated for over thirty years, finally leading to a settlement in which former marines received compensation for their diseases.

Preface

I wrote *The School Poisoning Tragedy in Caledonia, Ohio* to tell the story of a contaminated school district grounds, where questions were asked and data points provided of Examining Historical Evidence, Expanding Community Action & Communication, Defining the Results and Exploring Legal Options that offer a template for Formerly Used Defense Sites, Superfund and other contaminated sites. My aim is that my research will lead us to ask the right questions and to demystify the environmental site investigation results from their technical and scientific jargon so that they can be properly understood and interpreted.

I propose a *Health Risk Reimbursement Program* for Superfund and Formerly Used Defense Sites to provide a fairness approach for those who can prove that their debilitating diseases were directly caused by the contaminants at those sites and that such a program be retroactively implemented for the afflicted graduates of the River Valley Local School District in Caledonia, Ohio.

ACKNOWLEDGEMENTS

I would like to first thank my wife, Pat, once again, for her patience and encouragement during my lengthy hours of research and writing. To the graduates of River Valley High School, who continue to call into question the liability of the federal government for allowing the sale of contaminated property to the school district and for the graduates' strength in enduring years of hardship and debilitating diseases among the community. A special thank-you to the Marion County Public Library, especially librarian Kim Hunsicker, who assisted me in locating archived documents related to environmental site inspections and contamination of the school district grounds. Once again, to the great researchers at the National Archives and Records Administration, without whose assistance I would have been unable to retrieve historical photos for this book. Another thank-you to Mark A. Walters of SandPrint Memoirs & Editing LLC for his editing skills with both this book and my previous books, *A Tribute to the 109th Evacuation Hospital Semi-Mobile Unit* and *World War II POW Camps in Ohio*. Appreciation also goes to Bryan Miller at OfficeMax in North Olmsted, Ohio, for assisting in framing the photos.

A special thank you to John Rodrigue and Abigail Fleming of Arcadia Publishing and The History Press, who took me through every phase of the publication process to ensure that we had a quality product that would be appreciated by the reader.

I am grateful for Patrick Patton, a River Valley High School graduate, for his courage in telling of the pain and suffering he has endured with a brain tumor, which he attributes to the contaminated grounds of the school district. In his mind, no one ever cared or listened to the River Valley graduates who suffered like him, often at the hands of money and power. He told me: "If someone else out there cares, I haven't heard about them yet." I hope my story shows that someone out there cares.

INTRODUCTION

Questions are still being asked today.

The Marion Engineer Depot (MED) was the largest depot of its kind, housed on 650 acres and known for storing and renovating heavy construction equipment, shipped food, supplies and military equipment during World War II. The loading section of the depot that was laid out in four avenues and six streets, comprising twenty-two miles of railroad track. The depot was also the receiver of a Japanese submarine.[3]

The importance of the Marion Engineer Depot to the World War II efforts cannot be overstated. In one month, it loaded 1,544 railroad cars and unloaded 1,231 cars of material and equipment, as records show in May 1945. The facility hit a peak tonnage in May 1943 with 44,000 tons of troop stock shipped to war zones while it received 23,000 tons of equipment and supplies. During the wartime effort, the depot employed 1,400 civilians and held 295 prisoners of war who met the labor demands of the depot.[4]

This was the bright spot in the heyday of the Marion Engineer Depot. The property changed hands in the early 1960s when the River Valley Local School District purchased 78 acres, and they were subsequently used over the years as a middle school, high school and athletic fields. In 1997, to the heartbreak of the graduates and parents, it was discovered that a portion of school property had been contaminated all this time. It was in 1997 when residents of the school district believed that the high rate of leukemia, non-Hodgkin's lymphoma and other forms of cancer afflicting the graduates was

Opposite, top: Loading section of the Marion Engineer Depot. *Courtesy of the NARA Photo: No. SC 339601, July 27, 1943.*

Opposite, bottom: Japanese submarine. *Courtesy of the NARA Photo: No. SC 339586, July 2, 1943.*

Above: Japanese submarine. *Courtesy of the NARA Photo: No SC 339587, July 2, 1943.*

possibly linked to the former military activities at the depot, especially in and around the athletic fields.

The ensuing chapters will trace the plight of the graduates and parents of the River Valley Local School District's quest to find answers to why there were such high leukemia and cancer rates, answers which at times were stonewalled by both government bureaucratic incompetence and possible cover-ups. Appendix A lists the primary events related to the contamination of the River Valley Local School District, starting with the opening of the Marion Engineer Depot in the 1940s and continuing to the present day, a timetable that helps us to see the chief events that occurred but also that lingering, painful questions still persist.

1
BACKGROUND

Leukemia News Stuns Marion—Death Rates Up 122% in 30 Years, State Study Finds.

The River Valley Local School District built its middle school, high school and athletic fields on seventy-eight acres of the former Marion Engineer Depot in the early 1960s, but it was not until the late 1990s that residents of the district began to link the high rates of leukemia, non-Hodgkin's lymphoma and other forms of cancer prevalent among the graduates to the possible activities of the former depot. The schools were built on the same area that was once used for heavy equipment and training, military artillery practice, materials storage, burial of construction debris and the burning of waste material and fuels. A radium-painted dime-sized aluminum disc was found just below the soil on the front lawn of the high school, a marker the army used to indicate the position of bridges and vehicles while troops were practicing nighttime operations. Other tests at the school site indicated the presence of volatile organic compounds in both the soil and perched groundwater on school property. Vats of chemical residues were buried in trenches covered with soil, over which the school built its athletic fields, unaware of the toxic danger. Once this hidden threat finally became known, it led to citizen outrage and concern for the health and welfare of the students.

Another grave concern was the discovery of arsenic in the drainage ditches along both sides of a former railroad spur where prisoners of war were dropped off during World War II upon their arrival at the Marion POW Camp. Apparently, it was here that the prisoners of war were sprayed with an arsenic-laden delousing agent upon exiting from railroad cars known to be treated with an arsenic-bearing fumigant still present at elevated levels fifty years later.

An article in the *Columbus Dispatch* of October 17, 1997, titled, "Leukemia News Stuns Marion—Death Rates Up 122% in 30 Years, State Study Finds," tracked the growth in cancer rates among the local population. The accompanying chart, Tracking a Killer, put together by the state health department, compares the number of leukemia deaths in three ten-year periods for the city and county of Marion. A total of 151 deaths occurred among the city of Marion's 34,075 residents and the county's 94,274 residents (counted separately from the city) between 1966 and 1995. The article notes that the death rate from leukemia in Ohio over the same period was only 8 percent.[5]

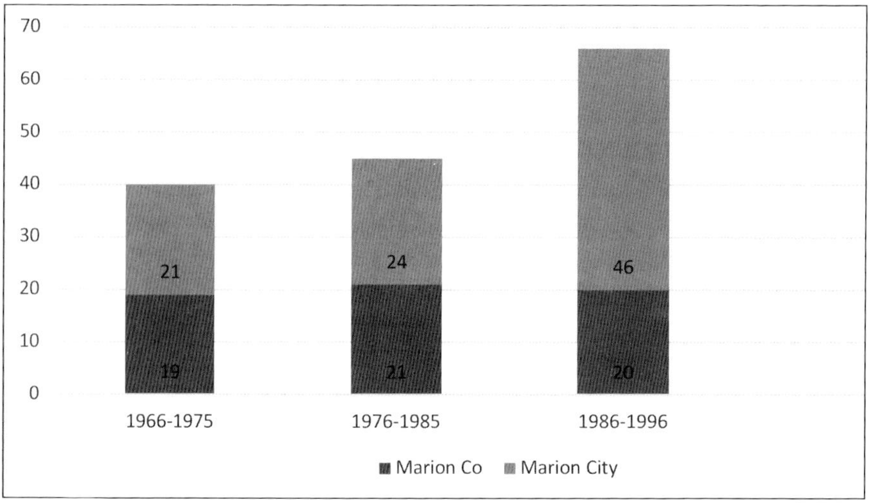

This alarming comparison led to concern across the region and state, especially among the families of the students who had graduated and those still attending the River Valley Local School District. Proactive measures were undertaken by the Concerned River Valley Families group, which periodically ran ads—one of which was "River Valley Graduates and Families, We Need Your Help"—stating that more than seventy cancers had been reported to the group.[6] The group's members feared that there may be some unreported cases.

The removal of the chemicals and hazardous waste at the school district site has cost $5.7 million as of a July 31, 2001 report.[7] In the end, approximately $8 million was spent on the school site cleanup and $35.7 million spent on the cleanup of the former Marion Engineer Depot grounds.[8] The schools were finally closed and rebuilt in a different location in 2003 at a cost of $53.4 million, including a $8,071,192 (21 percent) match on the Ohio Facilities Construction Commission, $30.3 million (79 percent) passage of a local bond issue and a $15 million contribution from the U.S. Army Corps of Engineers for relocation purposes.[9]

Concerns over the contamination of the River Valley Local School District reached the Ohio statehouse, where Governor Robert A. Taft stated that "after at least six months of discussions officials concluded that building a new campus would be cheaper than cleaning up the site enough for schools to stay. The soils have been investigated; they are clearly contaminated and the school has already lost a significant part of its property for ballfields and play areas."[10] After Governor Taft got involved, the Ohio 123rd General Assembly followed and wrote in the Amended Substitute Bill 282 line item 200–580: River Valley School Environmental Issues for Fiscal Year 2000 (July 1, 1999–June 31, 2000) in the amount of $350,000, which "Shall be used to assist costs arising from environmental assessment and cleanup of potential environmental contamination of school facilities of the River Valley Local School District."[11] The environmental issues assessment, funded by the Ohio General Assembly, led to a section in Amended Substitute Senate Bill 272 for Fiscal Year 2001 (July 1, 2000–June 31, 2001) that authorized the Ohio School Facilities Commission to provide assistance under the Exceptional Needs Pilot Program allowing school districts not on the priority list to qualify through the new Extreme Environmental Contamination of Classroom Facilities program. This section provided for a complete districtwide project instead of merely replacing a contaminated facility and for using any federal restitution as a means of reducing the bonded indebtedness that it would take to complete the project.[12]

This legislation that allocated funds to examine school environmental issues for a particular school district, including the River Valley Local School District, and the creation of the Extreme Environmental Contamination of Classroom Facilities section in the school facilities program was certainly an acknowledgement by the Ohio General Assembly that there were environmental issues on the River Valley Local School District campus that may have been negatively affecting the health and welfare of the students. Today, the vacated school buildings are being used as private warehousing facilities, with the requirement that the sites are to be tested every five years for contamination based on the remediation that was done to those sites.

The cleanup of the former Marion Engineer Depot was done through the Formerly Used Defense Sites program under the categories of CON/HTRW (Containerized Hazardous Toxic and/or Radioactive Waste) and HTRW (Hazardous, Toxic and/or Radioactive Waste), programs that comport with a highly toxic area and the cleanup of radioactive waste matching the activities at the former depot and the seventy-eight acres purchased by the school district.[13]

OBSERVATIONS

The school district was placed on the Ohio Facilities Construction Commission priority list because of extreme environmental contamination at the middle school, high school and athletic field sites. This program, because of the River Valley Local School District environmental issues, set the stage for other Ohio school districts to use it as they sought to move from contaminated sites. The Extreme Environmental Contamination program became a model that could be imitated by other states when encountering contaminated school sites.

Under the Formerly Used Defense Sites program, the former Marion Engineer Depot was classified as a highly toxic site with a possible correlation of the toxicity to the school district sites, yet health officials were not able to determine whether the contaminated soil contributed to the various forms of cancer that afflicted the graduates.

In a May 31, 2017 article in the *Columbus Dispatch* titled "Questions Linger Decades after River Valley Cleanup," reporter Dean Narciso encapsulated what many were feeling in the community after years of enduring the toxicity: "Today many of those who fought for more studies, accountability and justice feel that their happy high school memories have been dampened by what was buried under their playing fields." One former student indicated that "years of cross country, track and cheerleading put her in almost daily contact with the often-soggy school grounds." Another graduate contends that public pressure stopped further investigations that may have revealed the possible link between the contaminated school site to the high rates of cancer among former students.[14]

It is clear from hearing the voices in the community that the River Valley Local School District tragedy is an unfortunate legacy that will remain with the community for many years to come. Let us move forward with a detailed account of the contamination of the school district sites and lay out a disease and leukemia cluster to show how deeply affected the health of the graduates of the school district was by this contamination.

2
A TRAIL OF RADIOACTIVE MATERIALS

We were told from the time we got there that we will not find any radioactive contamination on the site anywhere. It wasn't directed as an observation it was directed as an order.

On April 17, 1947, a radiation survey was done at the Marion Engineer Depot of several thousand stored metascopes, also known as sniper scopes, that were retrieved from various war zones. These metascopes were used as night vision devices and contained radioactive radium in the foil. Three different night vison devices may have been stored at the depot. The first night vision device attached to a rifle, the second attached to a submachine gun and the third to a handheld device.[15]

The workers in the storage area wore film badges to monitor the intensity of the radiation and to limit the amount of time they could be in the storage facilities. One notation in the survey indicated, "Work involving the carrying of cases into the building from outside or vice versa, should be limited to an hour. If it is necessary to station a man inside the building, this time should be limited to fifteen minutes."[16]

The workers assigned to the buildings followed a tolerance level of 0.1 roentgens (unit of radiation exposure) per day or 0.7 roentgens per week. However, the limit of 0.7 roentgens per week could at times be received all in one day. A new guideline was developed that mandated that individuals should not stay inside longer than fifteen minutes. The results of the measures

Night vision device attached to a rifle. *Courtesy of the NARA Photo: No. SC-239417, January 1, 1946.*

Night vision device attached to a submachine gun. *Courtesy of the NARA Photo: No. SC-261501, November 12, 1946.*

Handheld night vision device. *Courtesy of the NARA Photo: No. SC 464440, June 29, 1954.*

found that twenty-three* out of the twenty-six locations within the buildings exceeded the recommended tolerance level of 0.1 roentgens per day:

RESULT OF MEASUREMENTS

<u>T 308 Building</u>

	Gamma Radiation <u>roentgens/8hrs.</u>
Inside doorway entrance outer door	.048
At the doorway along centerline of room, 4-foot-high cases on both sides	0.3*
4 feet inside	0.98*
8 feet inside	0.75*
12 feet inside	1.4*
End wall	1.6*
5 feet high against surface of case which was stacked in storage	0.5*
Against case surface 5.5-cubic-foot box	0.75*
" "	2.3*

<u>T 509 Building</u>	
At surface of lead box containing foils	0.25*
At doorway entrance	0.05
10 feet in from door along center of aisle, cases on left only	0.17*
20 feet in from cases on both sides	0.17*
Center of the cases of silica gel on the right	0.17*
10 feet from end, Samson cases on both sides	2.55*
At surface of cases (stacked)	0.63*
Electronic Samson	3.8*

<u>Miscellaneous Results</u>	
At surface of small opened carboard carton containing twelve metascopes	0.8*
One single Samson metascope at surface	0.6*
One single electronic metascope at surface	0.23*
One case in outside room, center surface	1.3*
6 feet away	0.5*
One case in outside room 12 feet away	0.3*
18 feet away	0.1
Outside T 308—at wall next to cases	0.26*
At fence line—north side	0.0016

Badge Results	Gamma Radiation <u>roentgens</u>
Badge 521	0.05
Badge 520	0.05

Above badges worn two and one-half hours, of which at least one hour was spent in the buildings housing metascopes.

Badge 522 was placed on 5.5-cubic-foot case for thirty minutes 0.05
Badge 523 was placed on 3-cubic-foot case for thirty minutes 0.18[17]

The radioactive metascopes were stored in specially designed buildings constructed of brick, lead-lined interior walls and a copper roof to house the radioactive materials. These materials were so potent that they would reduce by half in fifty years.[18] As radium decays over time it increases the amount of gamma radiation emitted to dangerous levels.

In an October 19, 1950 inspection of two buildings that stored metascopes, it was found that there were approximately one thousand unfinished instruments. Testing of the metascopes discovered that there was a decay of radium, and the radon had leaked through the plated coating on the charging foil and contaminated the phosphor (synthetic fluorescent).[19] The radioactive metascopes where removed from the buildings and relocated somewhere between 1955 and 1956. Could they have been dumped on the Marion Engineer Depot grounds?

By the end of World War II, the Marion Engineer Depot was storing fifty-eight thousand radium markers, which were eventually transferred to the Aberdeen Proving Grounds in Maryland.[20] It was said that some of radium-painted markers may have been lost around the site because they were not a controlled item, along with the radioactive pellets used to calibrate equipment and train radiological teams.[21] On the school district site during the 1950s, chemical, biological and radiological training occurred that involved gamma capsules being scattered in order to practice locating the capsules via Geiger counters and other radiation detection devices.[22] Once again, the questions are what kind of affect did this activity have on the grounds of the school district and were gamma capsules left behind and later buried in the soil of the school district sites?

A former depot worker who was assigned to Building 517 recalled that the building was marked with signs and symbols for radiation.

Loren Gast, who worked in Building 301 at the depot, remarked: "I never saw anyone around the building unless they were there strictly for business. It was off limits. It was kind of spooky." Gast said that before anyone entered the building, they had to put on radiation detection badges, and he did not know much about radiation because it was taboo to discuss it. Six months after he left the depot, he developed cancer and eventually had his left breast removed.[23]

Building marked with signs and symbols for radiation. *Courtesy of the Wikimedia Commons.*

According to Charles and Delpha Mosher's book *The Scioto Ordnance Plant and the Marion Engineer Depot of Marion, Ohio: A Profile of Forty Years*, the depot stored boxes of metals with protective wrappings designated for the

Practice locating the gamma capsules via a Geiger counter. *Courtesy of the NARA Photo: No. 111-SC- 427254, July 25, 1953.*

Manhattan Project for developing the atomic bomb.[24] These boxes may have been shipped to a Los Alamos laboratory.[25]

Did these metals leave behind a radioactive residue that could have permeated the soil?

An alarm bell went off in September 1997 when a dime-sized disc that was radium coated was found in the front yard of the River Valley High School.[26]

Los Alamos Scientific laboratory. *Courtesy of the NARA Photo: No. 306-PS-54-8361, June 11, 1954.*

Luminous marker disk.
Courtesy of the Battle Detective.

The disc gave off fifty-eight millirems of radiation, a measurement that the U.S. Army Corps of Engineers said was low but needed to be removed. The U.S. Army Corps of Engineers believed that the disc was a glow-in-the-dark marker used to identify soldiers or aircraft carrier decks at night.[27] The discovery prompted removal of the disc and further investigation by the U.S. Army Corps of Engineers of both the school grounds and the surrounding areas that made up the former Marion Engineer Depot.

Jed Ball, who was hired to do a radiation survey of the school grounds in 1998, claimed that his work showed fairly high levels of radiation in the surrounding areas; however, his findings were averaged with lower readings so that the results appeared normal. Ball essentially believed that his measurement of the radiation levels was doctored and that supervisors provided altered reports for him and fellow supervisors to sign. He stated: "We were told from the time we got there that we will not find any radioactive contamination on this site anywhere. It wasn't directed as an observation; it was directed as an order."[28] "They just make them up if they didn't like what we got."[29] His recommendations were ultimately rejected; other independent consultants were hired and found nothing wrong.

RATIONALE FOR RADIATION EXPOSURE

Cancers caused by radiation typically do not manifest until years after the exposure. Does this match the profile of the River Valley High School graduates who developed cancer? Were their cancers related to radium or other radiation contamination left behind by the former Marion Engineer Depot, which had a history of storing radium markers, atomic bomb metals and metascopes in lead-lined building walls and chemical, biological and radiological training? The radiation exposure pathways occur through skin contact, the inhalation of radioactive gases or eating and drinking substances that contain radioactive elements.[30]

The potential for external exposure pathways in which people can encounter radioactive pollutants are airborne, dissolved in the land and water and ingested, which certainly fits with the evidentiary pattern at the River Valley Local School District grounds.[31]

Non-ionizing and ionizing radiation emit unstable gamma rays as they decay over time and, through exposure, damage living tissue by changing the cell structure and damaging DNA. Since cancer from radiation often

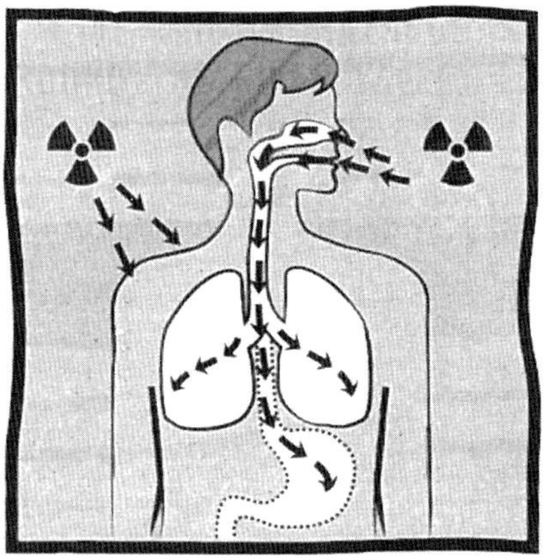

Left: Exposure pathways for radioactive elements. *Courtesy of the EPA, June 20, 2017.*

Below: Airborne external exposure pathways of radioactive pollutants that were airborne, dissolved and ingested. *Courtesy of the EPA, June 20, 2017.*

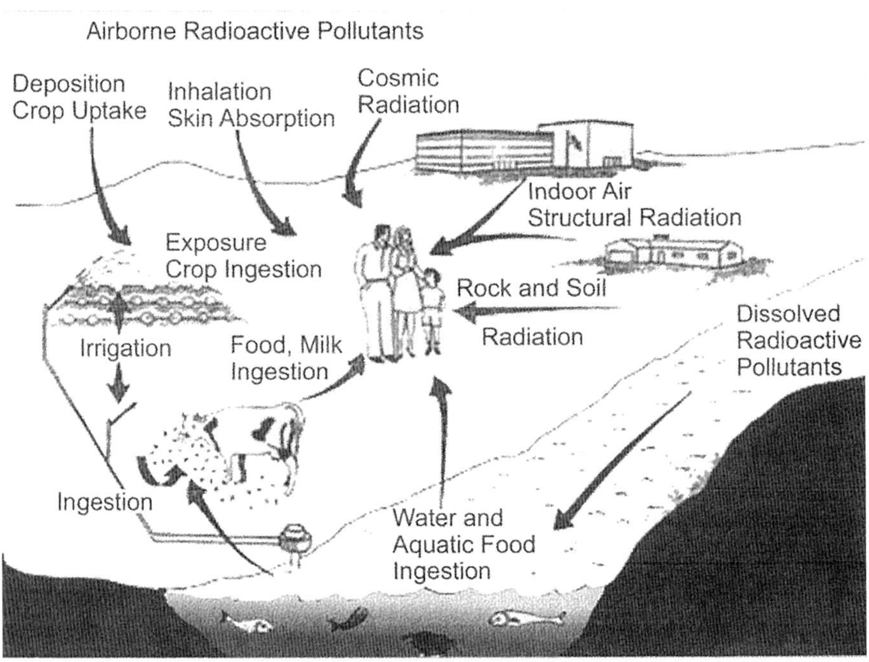

does not appear until years after the exposure, this may reflect that radiation sources were still strong and the decaying effect of released radiation formed other elements that were radioactive when the exposure occurred.[32]

Perhaps by the time the environmental site inspection reports were completed in the late 1990s and early 2000s, the strength of decaying elements of radiation were minimal and thus did not raise a red flag about radiation levels on the school grounds. The history of the former Marion

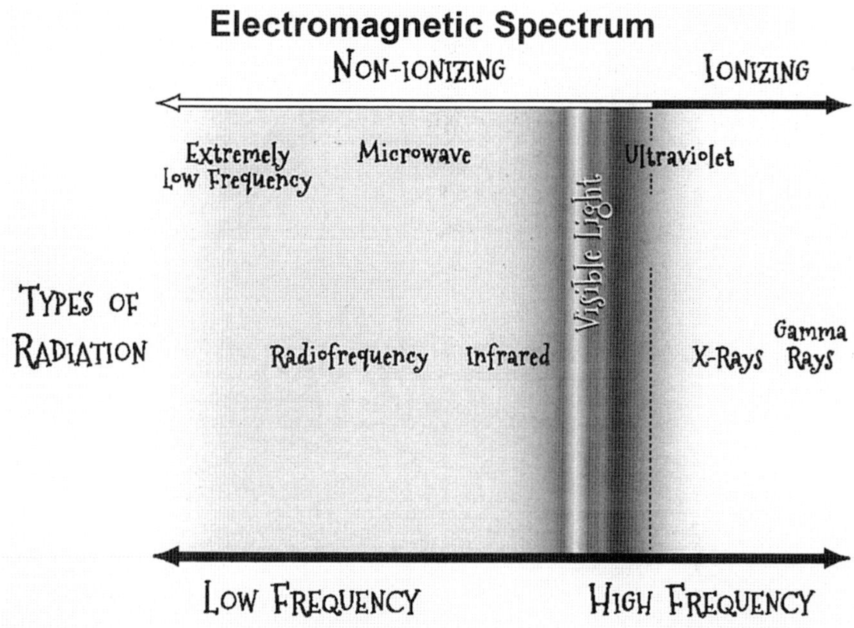

Non-ionizing and ionizing radiation. *Courtesy of the EPA, June 20, 2017.*

Engineer Depot gives reasonable cause to surmise that the River Valley Local School District's suspected leukemia cluster and other cancers of the graduates may have been caused by exposure to radioactive sources that had decayed over time and had pathways that permeated the school grounds similar as to what is seen in the Airborne Radioactive Pollutants diagram.

OBSERVATIONS

A Trail of Radioactive Chemicals portrays a history of possible radiation exposure for the employees at the former Marion Engineer Depot and the staff and students of the River Valley Local School District. It leaves behind a mystery of where the radioactive materials may have been dumped in the mid-1950s and raises the question of whether they were dumped on the depot grounds. Were some of the fifty-eight thousand radium markers left behind and disposed of somewhere on the depot site, where the schools and athletic fields were later built?

The environmental site inspection reports state that there were no traces of radiation left behind at the former Marion Engineer Depot, yet consultant Jed Ball, a nuclear health physics technician, reveals that he saw readings higher than expected—and what about the radiation badge worn by an investigator at the trenching site that had registered a significant amount of radiation that was left unexplained?[33] Furthermore, Patrick Patton, a graduate of River Valley High School, states that his father, who worked at the depot, told him that there was a lot of radioactive waste dumped into the dirt and then covered up where the schools were later built.[34] Given this direct evidence from the local community, it is hard to refute the potential for radiation exposure pathways over time affecting the staff, students and families of the school district.

3
TOXIC DILEMMA

River Valley property has the presence of multiple chemicals some of which amounts exceed acceptable by the U.S. EPA.

The River Valley Local School District faced a toxic dilemma with the amount of contamination that was found on the school grounds and the effect that it may have had on the staff and students of the school district.[35]

Dr. Bruce Molholt, a toxicologist and project consultant, believed that the schools should have been closed and the students transferred to different locations until the results of investigations determined whether the students were at risk. Dr. Molholt's survey results were rejected because officials at the Ohio Department of Health and the Ohio Environmental Protection Agency (EPA) believed that he mixed cancer data with environmental data. During his career, Dr. Molholt was called into numerous hazardous waste investigations to provide his expertise, and he often testified in court as an expert witness, yet his work was rejected because it deviated from the narrow parameters established by two Ohio bureaucratic agencies. Along with Dr. Moholt, environmental activists also pushed to have the schools closed and the students relocated to a safe environment.

The district's toxic dilemma was further exacerbated when Jed Ball claimed that his radiation survey was doctored so the results appeared normal. The district, as the environmental site investigations proceeded, experienced highs and lows, particularly anger and frustration from the

Workers digging up the soil of the River Valley High School checking to determine if hazardous materials are buried there. *Courtesy of the* Columbus Dispatch, *May 31, 2017.*

citizens, especially at the stonewalling by the government agencies at both the state and federal levels. Another complication was that the school district experienced a vacuum of consistent leadership, with three superintendents, one interim from 1997 to 2003, and three school board members that were not reelected.

The toxic dilemma started in August 1997 when military officials found a small source of radiation buried in the front yard of the high school, in addition to elevated levels of benzo[a]pyrene on the football practice fields. The material found on the fields was a heavy petroleum compound often found in asphalt, charcoal and roofing shingles.[36] The soil samples taken from the playing fields in November 1997 revealed high levels of polyaromatic hydrocarbons buried six to eight feet below the surface, which included benzo[a]pyrene, a chemical linked to cancer in humans after a lengthy and high-dose exposure.[37]

The athletic fields—which former depot employees described as a dumping site of fluids that more than likely included trichloroethylene (TCE), a fluid used to clean engine parts, all placed in trenches and burned—were roped off after the presence of chemicals was discovered.[38] At this time, the use of about one-fourth of the district's land, including five athletic fields, was restricted after chemicals were found to be linked to other forms of cancer.[39] In October 1999, it was reaffirmed that chemical contaminants and toxic metals were found on the athletic fields.[40] In addition, samples showing elevated levels of lead and trihalomethanes, a byproduct of the chlorination process, were discovered at the drinking fountains near the concession stands.[41]

Robert Ferguson, who worked at the depot from 1948 to 1959, recalled that an area at the east end of the property was known as the "pits" because automotive parts, spools of cable, paint strippers and solvents were either burned here or bulldozed and covered up. Bill Livingston, who worked at the depot in summers as a teenager, remembered hauling waste materials to the pit, where they were often doused with fuels. Livingston recalled, "That place always seemed to be on fire." Another chilling story was relayed by Ralph Hill Jr., who saw men in what he described as moon suits coming into the family home with Geiger counters to remove furniture his father had touched, in addition to taking his father's clothes—all without explanation.[42]

PCB, a potential cancer-causing agent, as well as two toxic metal detections for lead and arsenic were found at the dump area. Bacteria from

Athletic fields that were roped off after finding the chemicals on site. *Courtesy of the* Columbus Dispatch, *May 31, 2017.*

One of many rusted-out barrels that were found on the eastern edge of the U.S. Army Reserve training center adjacent to the River Valley Local School District property. *Courtesy of the* Columbus Dispatch, *May 31, 2017.*

human and animal waste was also discovered at the location of the former POW camp, which officials surmise may have been from the Marion World War II camp latrine. At the drinking fountains near the concession stands, samples showed elevated levels of lead and trihalomethanes, a byproduct of the chlorination process.[43]

Many rusted-out barrels found on the eastern edge of the U.S. Army Reserve training center adjacent to the River Valley Local School District property leaked carcinogens of chromium, trichloroethylene and polychlorinated biphenyls. A prominent hazardous waste sign warns visitors that the training center is an off-limits toxic waste site.[44]

Army Reserve lieutenant Jacob Tressler, who had trained at the site and later died from leukemia, believed at the time he was undergoing treatment that his illness was connected to his military service at the training center, but there was never a connection made to his illness and the toxic chemicals located there.[45]

DISPOSAL/DUMP SITE LOCATIONS

The disposal and dump sites housed rusted barrels leaking carcinogens at the reserve center, and bacteria from animal and human waste was discovered close to the school grounds at the former POW camp.[46] These discoveries added to the envelopment of the school grounds in carcinogens and toxic waste that may have contributed to a suspected leukemia cluster and other forms of cancer that afflicted the River Valley High School graduates. It is eye-opening to see how many dump sites were on the U.S. Army site where Lieutenant Jacob Tressler and other reservists unknowingly trained on contaminated grounds, exposure that may have led to the lieutenant's death.

It should be noted that $2 million was spent on removing eleven thousand tons of contaminated soil from the U.S. Army Reserve training site and that the Ohio EPA placed a restriction on the cleaned-up area that maintained that it was safe to be on the land so long as it was less than eight hours a day and no more than 250 days a year. A deed restriction was also placed on the land that prohibited permanent housing from ever being built on the training site.[47] These restrictions raise the question, Was the property really safe for the reservists to train on?

TABLE 1: DISPOSAL SITES NEAR THE MIDDLE AND HIGH SCHOOL LOCATIONS

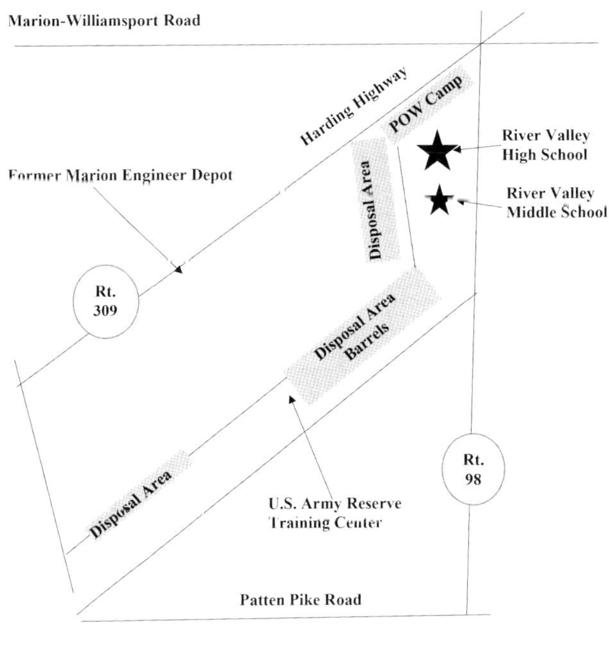

Investigators focused most of their efforts on the practice football field and baseball diamonds, where cancer-causing chemicals were detected. The U.S. Army Corps of Engineers confirmed that the area was used as a dump site. The fields were roped off in February 1998.[48]

Investigators also found chemicals linked to leukemia and other forms of cancer in quantities hundreds of times above acceptable levels in the soil just outside a chain-link fence surrounding the River Valley Local School District grounds. The chemicals were said to be "imminent threats" to humans because of the high potential that they could be ingested, touched or inhaled at the former military dump. The report found 310 times the acceptable amount of the chemical trichloroethylene (TCE) in soil two and a half feet deep in the dump nearest the school property.[49]

A photo from 1957 offered new clues: "River Valley schools could be resting on an old dump. A portion of the land on which River Valley Middle School sits what could have been used as a dump site. Acres of the scarred earth shown in the picture is in the same areas as the school, the anomaly areas and the Reserves dump site."[50]

A chain-link fence replacing a five-acre roped-off area was erected around the contaminated athletic fields at the River Valley campus after tests revealed more extensive pollution than was previously known, including benzo[a]pyrene, a pollutant commonly known to cause cancer at high doses, which was discovered in surface soils at levels higher than federal regulation requirements.

ASBESTOS HIDDEN ON SITE

Another carcinogen handled by the Marion Engineer Depot was asbestos, with as much as four thousand tons stored at the depot and eventually removed due to its deteriorating condition, according to Charles and Delpha Moser's book *The Scioto Ordnance Plant and the Marion Engineer Depot of Marion, Ohio: A Profile of Forty Years*. The authors maintain that it was never revealed that asbestos was being stored at the depot until, after extensive research, they discovered that it was essentially hidden from public knowledge.[51] The question is asked: Did some of the four thousand tons of deteriorating asbestos reach the school district grounds and add to the extensive contamination of the school sites?

SCIOTO ORDNANCE PLANT WASTELAND

Adjacent to the Marion Engineer Depot was the Scioto Ordnance Plant, which produced 395 cluster bombs per day, including 38 ten-pounders, each of which contained 15,000 small bombs. The cluster bombs ejected small submissions designed to kill personnel and destroy vehicles. The goop bomb was an incendiary oil- or gasoline-based bomb that was dropped in bombing raids over Germany and Tokyo. The plant employed 2,900 people who were responsible for manufacturing approximately 15.5 million fuses, 380,000 boosters, over 2.3 million primers, 2.9 million .50-caliber rounds, 100,000 five-hundred-pound incendiary munitions and even some ten-pound incendiary cluster munitions.[52]

The Scioto Ordnance Plant left behind toxic waste, with portions of the plant fenced off from public access. Lagoons or dumps have been discovered laden with concrete, bricks and burned materials on top of M-74 bomblets and incendiary petroleum-based fuels, including burned napalm. Over $11.2 million was spent on cleanup, after which a Defense Environmental Programs Annual Report to Congress for FY 2014 concluded with strong

Environmental cleanup of the Scioto Ordnance Plant. *Courtesy of the USACE, February 12, 2013.*

likelihood that civilian workers, army personnel and the prisoners of war had most likely been exposed to toxic working conditions.[53] One other account of the toxicity prevalent at the area comes from Norman Lord, who in an interview with EPA officials recalled that when he was a child he was told not to go into the quarry at the depot because it was toxic and that there were dead animals scattered throughout the site at any time.[54]

The question remains: Did the Scioto Ordnance Plant contribute to a suspected leukemia or cancer cluster in Marion County? The toxic waste cleanup of the site was equivalent to what was found at the former Marion Engineer Depot and the River Valley Local School District grounds.[55] It appears that the River Valley Local School District purchased property that was not subject to strict environmental oversight, a failing that may in turn have led to students suffering from cancer and other serious health problems.

OBSERVATIONS

There is a strong likelihood that students may have been exposed to hazardous waste left behind by the former Marion Engineer Depot and perhaps indirectly from the Scioto Ordnance Plant, where debris was transported and dumped on the depot grounds.

It is shocking how quickly the work of experts, such as Dr. Bruce Molholt and Jed Ball, was rebuffed and their findings rejected. The knowledge that the athletic fields were on the site of an old toxic waste dump and the possibility that the middle school was built on part of that dump was a missed opportunity, a leverage point for moving students off campus immediately instead of waiting six years for new schools to be built. Such a delay seems highly questionable, given the kind and scope of toxic contamination found on the school grounds.

The call to immediately have the students relocated to a safer off-site location while the new schools were being completed is adamantly expressed in a March 9, 2001 letter written by William A. Burga, the president of the Ohio AFL-CIO, to school superintendent Tom Shade. Burga's letter reads as follows:

> *Our organization represents over 800,000 workers and families in the state of Ohio, including several thousand in Marion County. Additionally, I am the father of four (4) River Valley High School graduates and*

stepfather of two (2) others. We are very concerned about River Valley High School still functioning at a location with the problems we have heard about as follows:

1. River Valley property has the presence of multiple chemicals some of which amounts exceed acceptable by the U.S. EPA.
2. Flawed investigations and investigation cover-ups, was disclosed by a federal judge.

We therefore believe it is important to move the River Valley students immediately and not expose them to any more health risks. Citizens have already voted for relocation, everyone knows of the contamination and they should not be at the facility waiting for a new school to be built.

It is further my understanding that Tri-Rivers Vocational School is willing to provide space for temporary classrooms in the interim period of construction of new River Valley schools.

Thank you for your consideration.[56]

Environmental cleanup of the River Valley Local School District. *Courtesy of the USACE, March 5, 2005.*

The magnitude of the environmental cleanup became more apparent once the schools were relocated and the removal began of twenty-seven thousand tons of contaminated soils with the cancer-causing agents arsenic, beryllium, and trichloroethylene from five acres of the River Valley Local School District's ball fields and athletic practice areas.[57]

Finally, the voices of Burga and those echoing his call were heard, that the students should be relocated immediately, not after six years, to ensure they were no longer subject to potential health risks that could affect them for the rest of their lives.

4
CARCINOGENS FOUND ON THE SCHOOL GROUNDS

We were running around in a toxic soup and didn't even know it.

The key to the River Valley Local School District property environmental site inspection reports centered on two reports: the Operable Unit 1 (OU1) Report of February 2000, which studied twenty acres used for disposal of construction debris and other toxic waste materials during the operation of the former Marion Engineer Depot, and the Expanded Site Inspection (ESI) Report of 2001, which surveyed fifty-eight acres and identified contaminants found outside of the OU1 area.[58]

EXPANDED SITE INSPECTION (ESI) REPORT (DECEMBER 2001)

Surface soil samples of TAL metals detected above site-specific background levels were chromium, iron, lead, manganese, mercury and zinc. In subsurface soil samples, the TAL metals that exceeded site-specific levels consisted of aluminum, chromium, copper, iron, lead, nickel and vanadium.[59]

Another finding that indicated detections above reporting limits included the semi-volatile organic compounds (SVOCs) benzo[a]pyrene, benzo[b]fluoranthene, benzo[k]fluoranthene, chrysene, fluoranthene, pentachlorophenol, phenanthrene, pyrene, bis(2-ethylhexyl) and phthalate,

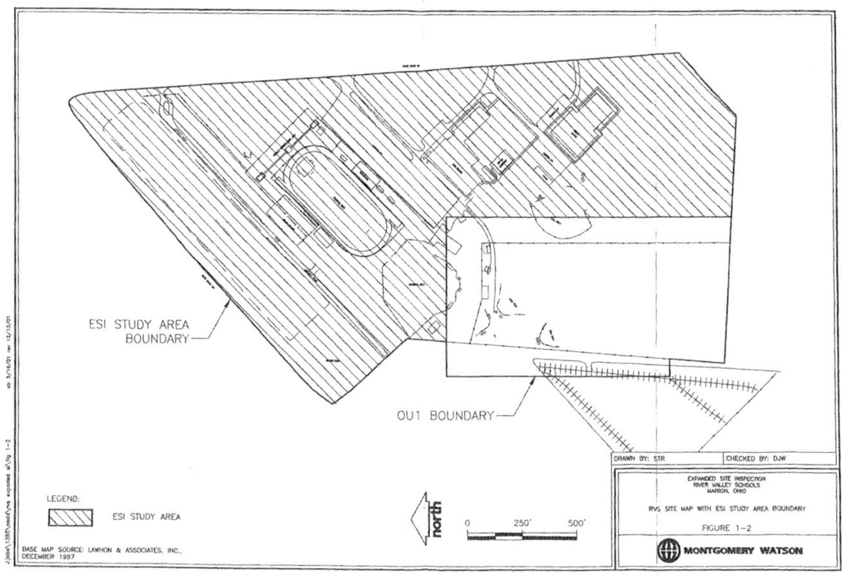

The environmental site inspection report areas, ESI Study Area and OU1. *Courtesy of the USACE, RVSD Property Expanded Site Inspection Report, December 2001.*

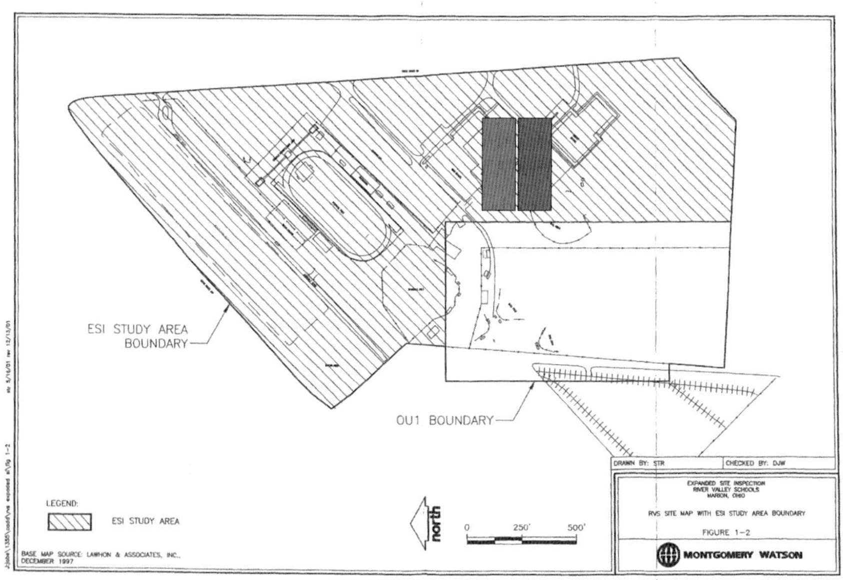

Site map indicating the locations of the SVOCs and TAL metals that were around the school buildings. *Courtesy of the author's interpretation.*

which were located mostly around the high school and middle school grounds.[60] The site map depicts the main locations where SVOCs and TAL metals were found around the school buildings.[61]

- Sixteen SVOCs above screening levels located on the middle and high school grounds.
- Fifteen TAL Metals above screening levels located on the middle and high school grounds.

Arsenic—a chemical that was used as a delousing agent and sprayed on German prisoners of war as they exited railroad cars and as a fumigant at the depot—was discovered on the school grounds, according to research uncovered while writing *World War II POW Camps in Ohio*. There were also several arsenic ditches located east of the school buildings along State Route 98.[62]

This begs the question as to what happened to the U.S. Army's recording-keeping of these events that used arsenic. It appears that only after a suspected leukemia cluster occurred in the school district did remedial action start to take place.

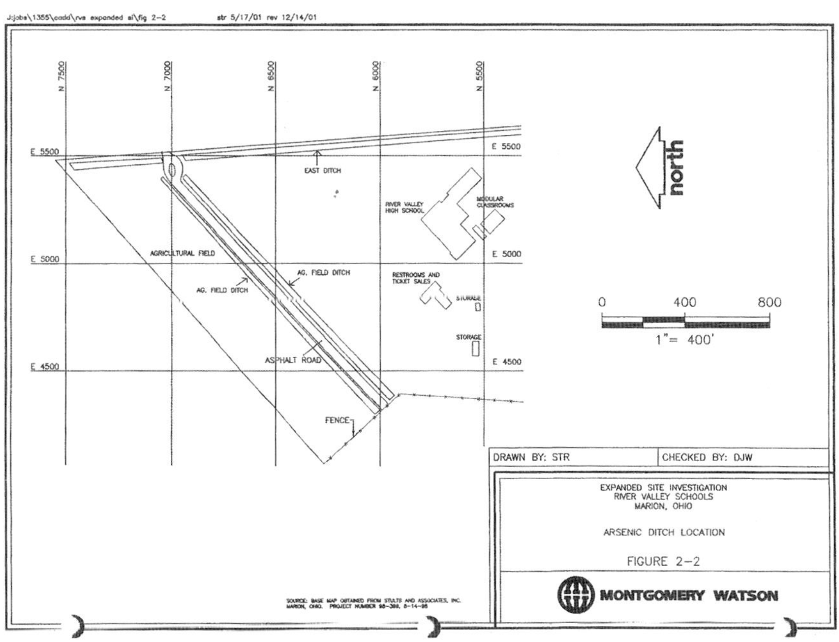

Arsenic ditches found east of the school buildings along State Route 98. *Courtesy of the USACE, RVSD Property Expanded Site Inspection Report, Figure 2-2, December 2001.*

Another review found that the cancer-causing agent benzo[a]pyrene was considered to be at low levels due to its presence found in roofing debris from buildings and in road dust, but the study did not reveal the contaminated locations: (1) the low area south of the middle school; (2) the front area of the middle school, between the building and parking lot; and (3) the front of the middle school, between the building and parking lot. In these areas, the arsenic found could not have been attributable to roofing debris or road dust, which thus raises the question whether these three locations had a negative effect on the environment of the middle school and high schools.[63]

The environmental site inspection report depicts the degradation pathway for the cancer-causing agents trichloroethane (TCE) and tetrachloroethene.

The report states: "Because of the limited amount of low level (below screening criteria) identified at the R.V.H.S. property (outside of OU1) biodegradation of contaminants cannot be determined."[64] This author's research shows that the pathway breaks down into vinyl chloride, with the outcome being the cancer-causing agent chloroethene, instead of chloroethane as depicted in the diagram. This research comes from the Agency for Toxic Substances and Disease Registry (ATSDR), which defines one of the synonyms for vinyl chloride as chloroethene with no mention of chloroethane as being a synonym. ATSDR also lists the chemical

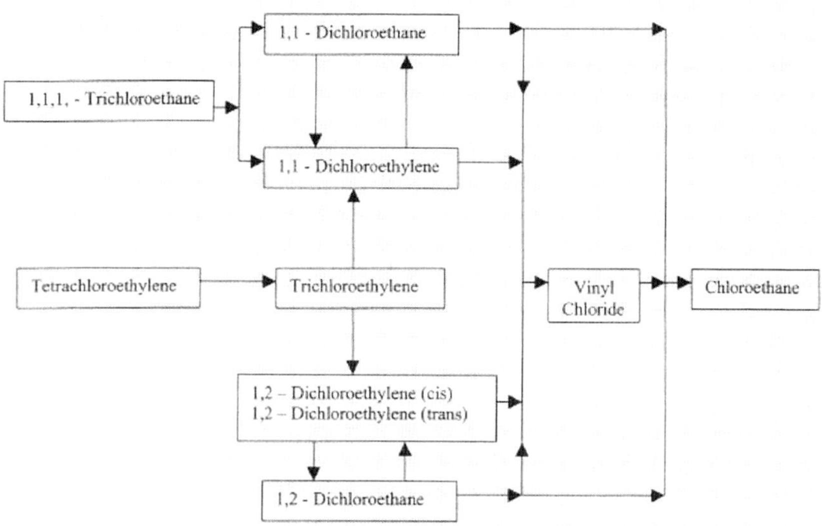

Degradation pathways for the cancer-causing agents of trichloroethane (TCE) and tetrachloroethene. *Courtesy of the USACE, RVSD Property Expanded Site Inspection Report, December 2001.*

formula for vinyl chloride (chloroethene) as C_2H_3Cl[65] while the formula for chloroethane is C_2H_5Cl,[66] which does not match the vinyl chloride (chloroethene) chemical formula and thus indicates that the depiction in the diagram is not an accurate representation of the pathway to vinyl chloride in the environmental site inspection report. Indeed, the International Agency for Research for Cancer (IARC) classifies chloroethane as not carcinogenic to humans[67] but classifies vinyl chloride (chloroethene) as carcinogenic to humans. This aligns with a far more negative effect on the graduates of the River Valley High School, who may have been at risk over the years of encountering the cancer-causing agent vinyl chloride (chloroethene), contrary to the government's misleading interpretation.[68]

This misinterpretation of the pathway as chloroethane instead of the carcinogen chloroethene was perhaps the reason that there were no red flags raised for more than forty years that biodegradation may have been a factor in causing a suspected cancer cluster and other cancers in the River Valley Local School District. Even though the diagram in the 2001 site inspection report shows the biodegradation led to chloroethane, the author's research indicates that the end product was, in fact, chloroethene, also known as vinyl chloride, which can cause liver, brain and lung cancer when exposure occurs through inhalation or drinking contaminated water.

Did the 2001 River Valley School Property Expanded Site Inspection Report get it wrong and show an inaccurate depiction of the biodegradation of trichloroethane and tetrachloroethylene at the River Valley Local School District sites, which misled the public into thinking that were no major contamination issues?

If one takes a thirty-thousand-foot view of the River Valley Local School District site that housed the middle school, high school and athletic fields, the evidence is clear that toxic chemicals at low and high levels permeated the soil throughout the campus, with pathways going 360 degrees across the seventy-eight-acre site.

OPERABLE UNIT 1 REPORT (FEBRUARY 2000)

Operable Unit 1 was the area used for the disposal and burning of construction debris and toxic waste materials at the former Marion Engineer Depot. The site encompasses the ball fields and a nature preserve located behind the school buildings in the southwest corner of the property.

Today, the area is occupied by a large car storage area surrounded by the former high school track.

The contamination of the area consisted of the following:

- Semivolatile organic compounds (SVOCs) detected in surface soils to one foot in depth over 3.3 acres; (1) metal concentrations in excess of both site-specific and regional backgrounds levels; (2) low-level carcinogen, polychlorinated biphenyl (PCBs) above the screening criteria; and (3) calculated dioxin/furan toxicity equivalency factors above screening criteria.
- Elevated concentrations of volatile organic compounds (VOCs), SVOCs and TAL metals above screening levels in surface and subsurface soils one to twenty feet in depth in a designated area of 6.3 acres that included both soil and subsurface waste materials disposed of during the operation of the former Marion Engineer Depot and PCBs and calculated dioxin/furan toxicity equivalency factors above screening criteria.
- Elevated concentrations of VOCs above screening criteria from one groundwater well.
- Metal concentrations above screening criteria detected in all wells in the Operable Unit 1 area.[69]

The VOCs found were primarily solvent-related compounds that included the carcinogens tetrachloroethylene, trichloroethene (TCE) and tetrachloroethene (PEC), plus trans-1, 2-Dichloroethene, a highly flammable colorless liquid. The TAL metals found were beryllium, chromium, cobalt, lead, manganese and zinc, all of which were detected above calculated background levels for surface soils. There were four samples of PCBs found, one of which was Aroclor 1260 PCB mixture.[70] The report states that "approximately the upper 2 to 4 feet of soil overlying the entire OU1 area has been disturbed as of result regrading, disposal activities, and/or construction equipment testing and operation. The disturbed soils may contain scattered and small amounts of construction debris (pieces of concrete, wood or asphalt) but generally appear to reworked in-place."[71] More than likely the contaminants that were found in the athletic fields were from the trenching, burning and burying of toxic waste such as paints, fuels, oils solvents and construction materials during the operation of the former Marion Engineer Depot. The site map for the

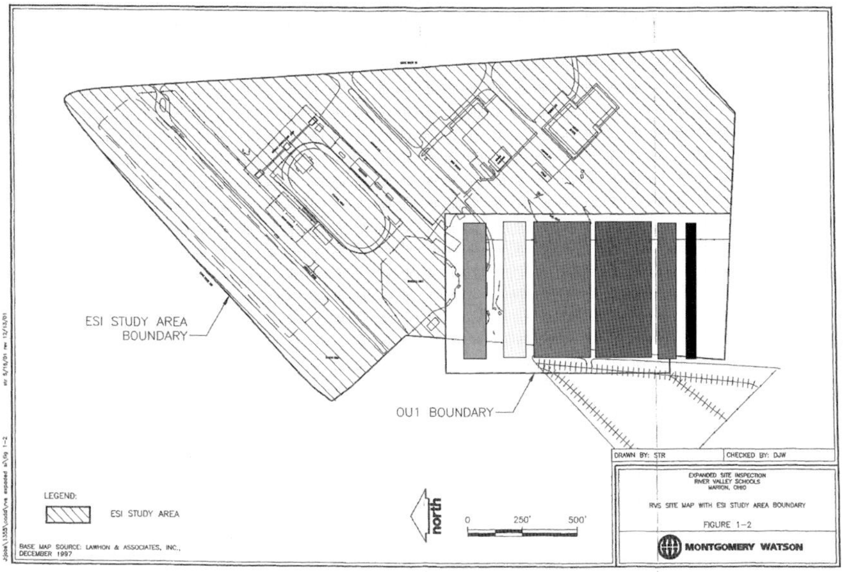

Site map for the OU1 area to illustrate the kind of contaminates that were found on the athletic fields. *Courtesy of the author's interpretation.*

OU1 area here illustrates the kind of contaminants that were found in that area of the school campus, namely the athletic field sites.[72]

- VOCs detected at twenty different locations.
- SVOCs detected at eleven different locations.
- SVOCs detected above USEPA IX preliminary remediation goals at thirty-five different locations.
- TAL metals detected at above the calculated background at forty-five locations.
- PCBs detected at three different locations.
- PCBs detected above USEPA IX preliminary remediation goals at one location.

One can see that under the former playing fields there was a mixture of VOCs, SVOCs, TAL metals and PCBs that formed a reservoir of toxic carcinogens upon which students unknowingly played for close to forty years. Melanie Serpa, who was diagnosed with Hodgkin's lymphoma in 1985 and remembers coming into contact with soggy school grounds, stated: "I remember there was a wet spot on the back field we ran through with wet socks and shoes. We used to joke that we were running through toxic sludge."[73] Kim Tolnar, who developed leukemia in 1993, reaffirmed Serpa's comment: "We were running around in a toxic soup and didn't even know it."[74]

The U.S. Army Corps of Engineers spent approximately $8 million to remove twenty-seven thousand tons of soil contaminated with cancer-causing agents from the athletic fields that may have led to the cancer of Kim Tolnar, Melanie Serpa and others who used the athletic field areas.[75] Removing twenty-seven thousand tons of contaminated soil equates to 1,125 standard dump truck runs containing twenty-four tons on each trip. This large-scale removal of toxic soil points to the enormity of the contamination at the athletic field and certainly must be recognized as a strong link to a suspected leukemia cluster and other forms of cancer suffered by the graduates of River Valley High School.

The National Toxicology Program's model of how trichlorethylene (TCE), a metal degreasing agent, is released into the air, water and soil and slowly breaks down, remaining in the environment for long periods of time, matches the scenario at the former Marion Engineer Depot and later

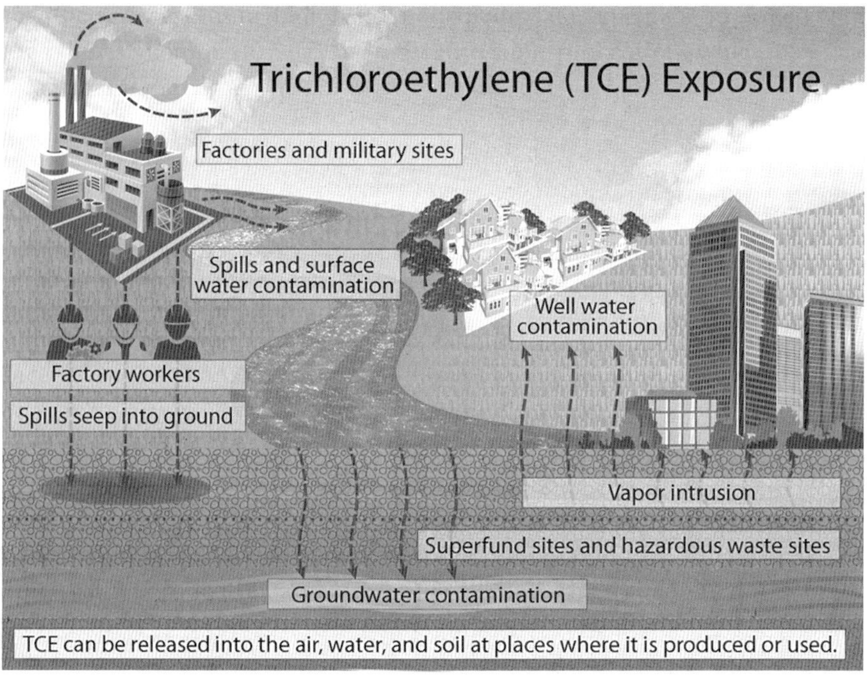

Exposure of trichlorethylene (TCE), which breaks down slowly and remains in the environment for long periods of time. *Courtesy of the National Toxicology Program.*

the River Valley Local School District athletic field areas.[76] The National Toxicology Program states: "TCE can move readily through soil and make its way into underground drinking water sources."[77]

This model paints a vivid picture of the various avenues of how the River Valley Local School District grounds may have been contaminated by TCE, carcinogenic to humans, that was left behind by the former Marion Engineer Depot in a toxic buildup that could have been a contributor to the debilitating diseases of the district's graduates who participated activities in and around the athletic fields.

MORE CARCINOGENIC EVIDENCE

The data shown in Table 2 lists the carcinogenic to humans agents that were found on the school grounds, the carcinogenic definition by the International Agency for Research on Cancer (IARC)[78] and the 2017 ranking of each of the carcinogens by the IARC for the Toxic Substances and Disease Registry Substance Priority List.[79] As the table shows, six of the top ten toxic substances ranked by the IARC were found at the school grounds.

TABLE 2: CARCINOGENIC TO HUMANS FOUND ON THE SCHOOL GROUNDS WITH TOXICITY RANKINGS

CARCINOGENIC TO HUMANS	DEFINITIONS
Arsenic – 1	Alloying additive for metals
Benzene – 6	Used as solvent and highly toxic
Benzo(a)pyrene – 8	Chemical that comes from substances that are burned such as oil and gas products
Beryllium – 43	Grayish metal when purified is used in nuclear weapons, reactors and x-rays
Chromium (VI) Compounds – 17	Alloying and plating element
Ionizing Radiation – 101	Rays given off by radioactive material
Lead Compounds – 2	Used in coverings of piping, etc.
Pentachlorophenol (PCP) & by-products – 54	Organochlorine compound used as pesticide and disinfectant

Carcinogenic to Humans	Definitions
Polychlorinated biphenyls (PCBs) – 5	Used as dielectric coolant fluids. Was discontinued in 1976 because of its toxicity and ecological damage done through water pollution
Trichloroethylene (TCE) – 16	Used as a metal degreasing solvent and banned in some states
Vinyl Chloride – 4	Used to make polyvinyl chloride (PVC) plastics and vinyl products

Table 3 further breaks down each of the carcinogens found on the school district grounds by its effect on different human organs with added toxicity rankings.[80] These numerous cancer-causing agents were suspected of secretly poisoning the personnel and students in a delayed reaction in which cancer and other diseases afflicted their organs later in life.

Table 3: Carcinogenic to Humans by Organ

Organ	Carcinogenic to Humans
Blood Disease (leukemia Lymphoma)	Benzene – 6 Pentachlorophenol (PCP) – 54 Trichloroethylene (TCE) – 16 Vinyl Chloride – 4
Brain	Vinyl Chloride – 4 Lead Compounds – 2
Colon	Arsenic – 1
Kidney	Arsenic – 1 Chromium (Hexavalent) – 17 Lead Compounds – 2 Pentachlorophenol (PCP) – 54 Trichloroethylene – 16
Immune System	Benzene – 6 Beryllium – 43 Chromium (Hexavalent) – 17 Polychlorinated Biphenyls (PCBs) – 5

ORGAN	CARCINOGENIC TO HUMANS
Liver	Arsenic – 1 Pentachlorophenol (PCP) – 54 Trichloroethylene (TCE) – 16 Vinyl Chloride – 4
Lung	Arsenic – 1 Beryllium – 43 Chromium (Hexavalent) – 17 Pentachlorophenol (PCP) – 54
Nervous System	Benzene – 6 Ionizing Radiation – 101 Pentachlorophenol (PCP) – 54 Polychlorinated Biphenyls (PCBs) – 5
Skin	Ionizing Radiation – 101

The identified carcinogens found on the school district grounds were possible links to the cancers that afflicted the graduates. The tables provide a vivid picture of the possible toxicity of the school grounds, especially when one considers the different carcinogens that fit with the site map areas of toxic contamination in and around the athletic fields.[81]

OBSERVATIONS

The initial conversations by the government agencies indicated that they believed the school district grounds were safe for students, but this notion was challenged by the findings in the Expanded Site Inspection Report and QU1 Report. How could the government agencies hold that given that a toxic waste dump made up part of the athletic field areas, the toxic waste at the U.S. Army Reserve facility adjacent to the school district and the high levels of radioactive materials stored at the former Marion Engineer Depot? Imagine once playing on athletic fields unaware that they were possibly contaminated with toxins and now living in fear that your cancer may reoccur or you may someday get cancer because government agencies failed to link the carcinogens that permeated the soil to the cancer afflicting the community?

One the most outlandish misrepresentations of facts was that the biodegradation of the cancer-causing agents of TCE and tetrachloroethene led to chloroethane, and not the carcinogen chloroethene (vinyl chloride), which leads one to believe that the science used in the 2001 site inspection report may very well have been flawed. Such was the conclusion reached by consultants Jed Ball and Dr. Bruce Moholt from the outset and whose research, tragically, was pushed aside.

5
COMMUNICATION EQUALS AWARENESS

Over 70 cancers have been reported to us thus far and we fear there may be many more.

The activism by graduates and parents of the River Valley Local School District whose loved ones developed leukemia and other forms of cancer is credited with pushing state and federal agencies to undertake environmental site investigations that found that the athletic fields were built on a toxic waste dump, in addition to the other toxic waste material discovered on the school district grounds left behind by the former Marion Engineer Depot. The Concerned River Valley Families group placed ads in the *Marion Star* as a means of gathering information from former graduates who had cancer, appealing to the community to come forward with other cancer cases in order to pressure government agencies to do in-depth studies of the school district sites.

RESTORATION ADVISORY BOARD

Because of the actions of the Concerned River Valley Families, the Restoration Advisory Board (RAB) was created in August 1998 and began holding monthly meetings to discuss the progress of the environmental site investigations of the River Valley Local School District, including the former Marion Engineer Depot and Scioto Ordnance Plant grounds, so

Table 4: The Concerned River Valley Families Ad

RIVER VALLEY GRADUATES AND FAMILIES
WE NEED YOUR HELP

The U.S. Army Corps of Engineers is about to begin the risk assessment of the River Valley High School Campus. We of the Concerned River Valley Families feel the risks have not been fully investigated. The newly discovered potential dumping site is evidence of this. We feel it is extremely important that any possible link between the Carcinogens already found on the site and the Cancer cases reported to the Concerned River Valley Families be a part of the Risk Assessment. Over 70 Cancers have been reported to us thus far and we fear there may be many more. Toxic Chemicals such as Trichloroethylene (TCE), Arsenic, Benzene, Vinyl Chloride, Chromium, Benzo(a)pyrene, PCBs and many other chemical contaminants have already been found on the school grounds. It has been our goal from the onset to ensure our children safety and to do this we must have as complete a list as possible for all Cancer cases among River Valley Graduates, to properly and fully assess the problem.................... Please help us.

IF YOU OR SOMEONE IN YOUR FAMILY GRADUATED FROM RIVER VALLEY HIGH SCHOOL AND HAD OR HAS ANY TYPE OF CANCER PLEASE CONTACT ONE OF THE FOLLOWING PERSONS

Robin Millard
Kent of Roxanna Krumanaker
Mike Griffith
Barry or Melanie Serpa

IF YOU HAVE CONTACTED US IN THE PAST CONTACT US AGAIN TO CONFIRM THE INFORMATION, WE HAVE IS CORRECT. THE IDENTITY OF ALL PERSONS CONTACTING OUR ORGANIZATION WILL BE HELD IN STRICTEST CONFIDENCE.

THIS SPACE SPONSORED BY THE CONCERNED RIVER VALLEY FAMILIES

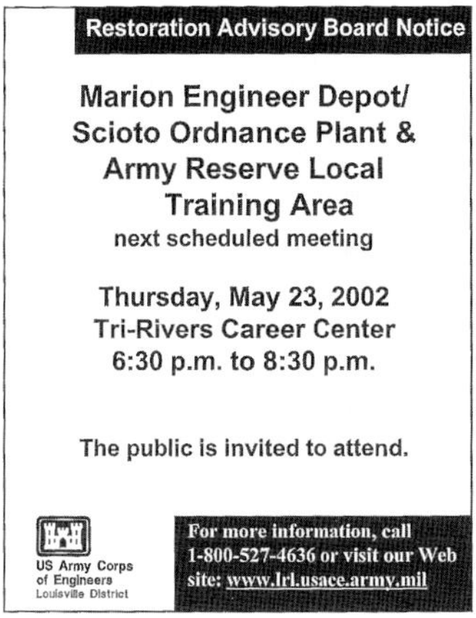

Restoration Advisory Board ad. *Courtesy of the* Marion Star, *February 8, 1999.*

that the community could provide input to decision-makers. There were twenty members and ten alternative members on RAB, with agency representation from the U.S. Army Corps of Engineers, Ohio EPA and Ohio Department of Health. A community co-chair was established with the project manager from the U.S. Army Corps of Engineers, and in 2001 a third co-chair was added with the U.S. Army Reserve, since the organization was going through a cleanup of toxic barrels found on reserve property.[82]

The RAB continued to hold meetings with presentations from various government agencies and consultants until its adjournment in July 2010. During the twelve years of meetings, there were over ninety reports presented at RAB gatherings. They were filed in an Information Repository at the Marion County Public Library in Marion, Ohio, and are available for the public to review.[83]

The RAB was able to provide ongoing public dialogue about the current environmental site investigations. Most important were the updates related to the River Valley Local School District grounds that were held in an open forum and allowed the public to interact with government officials. This format had been missing prior to the formation of the board. RAB ceased to exist in 2010 due to the absence of members for over three years, but the cleanup continued and was finalized in 2012 after fourteen years of investigations.

OTHERS SPEAK OUT

The American Cancer Society's Marion branch became involved by developing forums on early detection and prevention of cancer, in addition to hosting discussions about such cancers as leukemia, breast, lung, prostate and colon, as well as other diseases. These kinds of forums helped Marion County residents gain a better understanding of these dreadful diseases and offered some practical measures on how to prevent cancer.[84]

The Ohio EPA finally ended its silence and began providing weekly informational updates on past and current activities, summaries of the week's activity and ongoing activities, documents which afterward were submitted to the Marion County Public Library's Information Repository. The Ohio EPA also continued to remind the public that they could meet and ask questions of RAB committee members prior to the regularly scheduled meetings.[85] This kind of communication certainly increased the awareness of the citizens in Marion County about the environmental site investigations of the school district, the former Marion Engineer Depot and Scioto Ordnance Plant, marking a complete turnaround in communication by the Ohio EPA that was previously inconsistent or lacking altogether.

In December 1998, Barry Serpa, whose wife was a 1983 River Valley High School graduate and twice contracted Hodgkin's disease, a cancer of the lymph nodes, asked the school board to consider relocating the middle school and high school students until the environmental site investigations were completed. Melanie Serpa said, "They should err on the conservative side of the children's health. They should temporarily relocate until there is conclusive documentation that there isn't a health risk."[86]

Tom Snyder, a school board member, took the initiative to remove his daughter from the high school and announced that she would be attending a neighboring school district. Pam Snyder, a cancer survivor, said: "I don't know if there's anything out there or not out there but I have had cancer, so would we be doubling-exposing her? If she were to come to me later, and she had cancer and asked, 'Why didn't you take me out?' What would I say?"[87] The main concern of the Snyder family was that their daughter had spent time on the fields that were close to the roped-off area considered ground zero for the contamination, and they did not want their daughter to be exposed anymore to the contaminated school grounds.

Local resident Paul Christian raised concerns about the possible contamination of the school grounds by writing a letter to the editor in the March 29, 1999 *Marion Star* titled "Better Safe Than Sorry," in which he writes, in part:

> *I would much rather take my child out of school and find out that I was worried over nothing than leave my child in school and find out in a more deadly way that I was wrong.*
>
> *I can remember in the past what we told that Agent Orange was no big deal. I can remember in the not so distant past when we were told that there*

> *was no such thing as Gulf War Syndrome, even though as a veteran of the Gulf War I know better. Who do you trust? I am not saying over-react. I am simply asking what is the price for under reacting?*[88]

The environmental site investigations of the school district spurred activists, including Lois Gibbs, executive director for the Center for Health, Environmental and Justice and known for her work in raising awareness at the Love Canal contaminated school grounds at Niagara Falls, New York. Gibbs and other environmental activist groups pushed for the closing of the River Valley schools in 1999 by lobbying for more action by the Ohio EPA.

The Ohio Citizen Action, an environmental group based out of Columbus, posted a notice for a vigil for students at River Valley schools to be held on Monday, August 21, 2000, outside the governor's mansion in Columbus. The group hoped the vigil would create a response from Governor Taft to close the schools and reassign the students to other schools in the area. The posting is found in Appendix B.[89]

In a June 2, 2000 letter to the executive director of the Ohio School Facilities Commission, Ohio EPA executive director Christopher Jones emphasized the need to relocate the River Valley Local School District campus and to reassure everyone in the interim that the students and staff would be protected from environmental contaminants because cleaner soils were used as a surface barrier preventing the contaminants from reaching school grounds. Jones indicated that once the students were relocated, the topsoil barrier would be removed to excavate the most contaminated subsurface soils, meaning buildings would need to be empty because of the potential for exposure to the contaminated soils.[90]

Up until the June 2, 2000 letter, it appears that students were possibly subjected to uncontrolled contamination, which very well could have contributed to higher cancer rates among the school district's graduates. The Ohio EPA executive director's recognition that subsurface soil environmental contaminants were indeed present seems to mark the first time that an agency administrator believed there was a serious problem and that the schools needed to be relocated as soon as possible. Such a conclusion was what parents and activists had been recommending all along in order to safeguard the health of the students.

OBSERVATIONS

It is evident that from the beginning of this crisis, government agencies were ill-prepared to deal with these contamination issues. The members of the Concerned River Valley Families action group deserve recognition for their continued push to advance the cause that the community was being negatively affected by the soil contamination—so much so that it led the Ohio EPA, Ohio Department of Health and U.S. Army Corps of Engineers to at last communicate the progress of the environmental site investigations. The culmination of this long and devoted effort was the statement by Ohio EPA executive director Christopher Jones that the relocation of the schools would be in the best interest of safeguarding the health and welfare of the students.

More progress came at the hands of the Restoration Advisory Board, which played a pivotal role in keeping the public aware of the findings of the environmental site investigations. The Ohio EPA's weekly updates also proved to be a positive communications strategy that led to cleaning up the school site. No less vocal were the other calls to action by environmental activists who pushed for closing the schools.

Initially, the communication about possible contamination of the school district grounds seemed inconsistent and lacking, but because of the efforts of dedicated parents' awareness it was much improved. There is, however, still today a lingering sense of unease from some, including the Serpas and Snyders, over the slowness in relocating the schools and a belief that an earlier closing of the schools would have helped to prevent the growth of new cancer cases.

6
THOSE WHO KNEW

Remembers gasoline, fuel oil, antifreeze and motor oil drained and buried/burned in a pit that was in the area of where the schools are now located.

Between January and March 2000, Ohio EPA officials conducted interviews of individuals who worked at the former Marion Engineer Depot. The former employees provided information about the toxic wastes that were buried and burned at the site of the River Valley Local School District and former Marion Engineer Depot.

The interviewees in this chapter consist of George Cerny (sheet metal and welder), Dwayne Everly (battery repairman), Robert Ferguson (deputy safety director), Loren Gast (crater and crater foreman), Harold Gibbons (care and preservation), Donald Gorrell (care and preservation), James Haycox (mechanic), Ralph Hill Jr. (civilian worker), John Johnson (MRVHS/H&H Excavating), Bill Livingston (summer worker), Norman Lord (care and preservation), Bob Lowry (paint and maintenance supervisor), Melvin McKnight (maintenance shop), Don Patterson (fire department), Bob Snader (surveyor), John Tincher (Grief Brother Construction and National Guard member) and Richard Trimmer (sandblasting monitor).[91]

WHAT KIND OF HAZARDOUS WASTES WERE DUMPED IN AND AROUND THE RIVER VALLEY LOCAL SCHOOL DISTRICT GROUNDS?

GEORGE CERNY: Tanks that were half filled with contaminated gas ("red gas") were dumped and burned about where the high school was built.

DWAYNE EVERLY: While working at MED as a battery repairman, Everly related that the old batteries that could not be refurbished were taken to a landfill on pallets and dumped. He noted that a lot of batteries were dumped in a landfill that was located at the east end of the depot, where the baseball diamonds are now located. Everly also indicated that other materials were dumped in the landfill, including old paint cans, solvents from cleaning parts and a material called Electrolux, which was used in replenishing the batteries.

ROBERT FERGUSON: The safety director at the MED remembered that most disposals took place in a large area on the east side of the depot, where the schools are now located.

LOREN GAST: Did not remember any dumping by the school but heard that they did disposals there.

HAROLD GIBBONS: Remembered gasoline, fuel oil, antifreeze and motor oil that was drained and buried or burned in a pit where the schools are now located.

BOB LOWRY: Found arsenic-contaminated soil near a ditch on State Route 309 and believed that he was exposed to it because he got stuck in the area while driving a truck. Later, Lowry was diagnosed with some type of disease associated with arsenic exposure. Also, there was an arsenic ditch that was found on the school grounds.

MELVIN MCKNIGHT: Remembered solvent waste drained into a grate with tiles toward the east and south of the school grounds.

DON PATTERSON: Remembered Bailey bridges being stored where the schools were located. Patterson recalled that the bridges were sandblasted and repainted and felt that the problem at the school occurred by the stuff being brought onto the school grounds while the school facilities were being built. The photos of the Bailey bridges bring to mind, after sandblasting and painting, the extent of the bridges and kind of debris that may have been left behind and possibly mixed into the soil during the construction of the school buildings and athletic fields. As a matter of fact, as the Marion Engineer Depot River Valley School Expanded Site Investigation report indicates, at the construction site of the schools: "Surface soil grading for construction of these structures may been re-distributed contaminants associated with the Army's activities, and

Bailey Bridge across the Rhine River. *Courtesy of the NARA Photo: No. SC 32911, April 18, 1945.*

additional anthropogenic contaminants may have been added to the site surface soil."[92] So more than likely from the get-go, the school sites were possibly contaminated with toxic waste that was dispersed into the school construction site soils.

The first photo is the Engineers of the 167th Engineer Combat Battalion constructing the first Bailey bridge across the Rhine River near Wesel, Germany, on March 26, 1945, where they were under fire continuously for twelve hours.[93] The second photo shows engineers with the 35th Engineer Combat Battalion constructing a Bailey bridge over the Nisqually River in Washington State on October 31, 1951.[94]

BOB SNADER: While during survey work on the Marion Engineer Depot grounds in the late 1960s or possibly the early 1970s, he observed four-to-five-foot-tall piles of silver-colored metallic ore-like material that was being cleaned up between the buildings and the school.

JOHN TINCHER: After working for Grief Brothers Construction in the early 1960s, Tincher remembered there was a disposal area behind the buildings

Bailey Bridge across the Nisqually River. *Courtesy of the NARA Photo: No. SC 327911, October 31, 1951.*

near the main entrance and solvents were dumped in a ditch. Drums were stored near the ditch, and contents were dumped in the ditch, which ran the length of the depot toward the area where the schools were located. The disposal of waste material was done every Friday.

Mr. Tincher said that during National Guard maneuvers, 105mm Howitzer cannons were used for practice on what is now the school property.[95] The question is: What kind of contaminants were left behind and were they sufficiently removed from the artillery practice site or instead mixed into the soil?

Mr. Tincher also remembered seeing railroad ties, trash, trees, lumber and other materials being burned in an area near the schools, another area near the schools that appeared to be dug up and never had grass growing in it and a ditch nearby that did not have vegetation and always had a solvent or tar odor that never dissipated.

A 105mm Howitzer practice range. *Courtesy of the NARA Photo: No.111-SC-251529, March 6, 1944.*

A 105mm shell in the breach preparatory for firing. *Courtesy of the NARA Photo: No. 111-SC-251530, October 31, 1951.*

Richard Trimmer: Remembered the sandblasting of Bailey bridges that was done in the open with the paint and debris falling on the ground. The sandblasting took place where the schools are now located. It should be noted that the Bailey bridges had radioactive reflective markers.

At What Other Locations of MED Was Hazardous Waste Dumped?

Dwayne Everly: Remembered that when he worked in base maintenance, they used solvents to clean parts and paint equipment, with the waste being dumped on site and paint chips from sandblasting Bailey bridges dumped into a landfill.

Robert Ferguson: Remembered that while he was safety director of the MED, most of the waste disposed of was from the maintenance building, which included the carcinogens trichloroethene (TCE), chloroethene (PCE) paint strippers and two preservatives—known as P1 and P2—which were petroleum-based products used for long-term storage of equipment. The preservative was spread on the equipment and wrapped with a preserving tape and then stored in the depot.

Ferguson also related that degreasing took place by heating TCE and lowering the equipment into a vat, with the waste ending up in a creek adjacent to the building. The creek ran the length of the depot and into an adjacent farm field. Sludge from the cleaning process was put into drums and the drums subsequently burned in pits with other debris such as wood, scrap lumber, gasoline and oil. He believed that the drums contained TCE. Ferguson also noted that waste from sandblasting the Bailey bridges was either buried of burned.

Loren Gast: Remembered there was a salvage yard where they dumped stuff into pits.

Harold Gibbons: Remembered the carcinogen TCE being used for degreasing, cleaning and taking preservation off parts was put into drums and dumped into pits with up to fifteen to twenty drums per month. Gibbons said that the word *trichlorethylene* was written on the drums.

Donald Gorrell: Recalled oil and gasoline being disposed of into burn pits that were forty feet in diameter and burning was done three times a week.

James Haycox: Saw paint and wood burned west of where the water tower was located.

RALPH HILL JR.: Remembered his father taking him along Route 309 near where the POW camp was located and saw materials burning in open pits.

JOHN JOHNSON: Cylinders were found at a manufacturing plant behind the depot, and they were removed along with contaminated soil. He heard that gallons upon gallons of cosmoline were buried on the MED grounds as the depot was closing.

BILL LIVINGSTON: Worked at the Scioto Ordnance Plant, where he hauled paint cans, lumber and other scrap to the MED, with the waste material being dumped, burned and buried in trenches and pits behind the depot. The waste material was covered up with three to four feet of soil. Livingston also noticed drums in pits ten to twelve feet deep that were dug by the U.S. Army Corps of Engineers. He believed the drums and waste contained carbon tetrachloride or chlorine because parts were cleaned with a solvent. He also asserted that various other substances were buried on the Army Reserve property.

NORMAN LORD: Remembers TCE being dumped east and south of the buildings at the end of the railroad tracks. They dumped between a half to a full load of fifty-five-gallon drums a week. In the cleaning process, they also used acid, paint stripper and Stoddard solvent for cleaning parts, which eventually was dumped into a pit about the size of a football field but rounder. The middle was about ten to thirty feet deep, and fill dirt was bulldozed over the top of it. They also burned paper and other debris in the salvage area. Lord did say that he was tested every six months for contamination and they never found anything wrong with him, but he felt that they would have never told him if there was something wrong.

BOB LOWRY: Remembers Xol waste paint thinner being dumped into trenches on the east side of the depot just south of the last road.

MELVIN MCKNIGHT: While working as security guard, McKnight would check on a waste pit to see if it was burning but only saw smoke from the pit. He also recalled a burning pit on the east side of the depot, plus another pit in the salvage area where the fire department oversaw the burning pits.

DON PATTERSON: Remembered there was a pit on the west side of the depot where factories working outside of the depot dumped various debris in the 1960s, mostly wood and paper. Patterson also remembered burning pits that contained oil, lumber and paper located on the southeast side of the depot.

JOHN TINCHER: He said that sand from the sandblasting operations of the Bailey bridges was taken and disposed of on what is now the U.S. Army Reserve property.

WHAT ABOUT RADIATION ON THE MARION ENGINEER DEPOT GROUNDS?

Radiation film badge. *Courtesy of the NARA Photo: No. 22118242, June 21, 1948.*

Robert Ferguson: Remembered the Army Environmental Health Lab from Bainbridge, Maryland, came and decontaminated Building 517 where the radioactive metascopes were stored. As safety director, Ferguson sat in on the MED chemical, biological and radiological (CBR) team meetings.

Loren Gast: Remembered packing and unpacking metascopes and wore a radiation badge[96] when he and others entered the building. Gast developed skin cancer, which led to a mastectomy on his left breast and a tumor on his colon.

Ralph Hill Jr: Remembered his father talking about "hot water" and a leaded room that was reportedly located at the MED. In the 1950s, there was a spill of hot water, and his father was part of the cleanup team that used a tow motor. At home, men in white suits came to the house with Geiger counters and removed furniture and clothing that his father had been in contact with. His father wore a badge at the MED, which Hill believed was used to detect radiation. Hill indicated that most of his brothers and sisters had diseases, including cancer.

Bob Lowry: Served as a member of the CBR team that monitored atmospheres throughout MED. Lowry remembered finding a small amount of radioactive contamination by the front gates. He also remembered there was a response to radiation leaking in the air from the metascope building, and when he opened the metascope building the instruments went off.

Melvin McKnight: Bailey bridges had radioactive reflective markers and were stored in separate buildings.

Don Patterson: Inspected every building except for two because they were not allowed on the premises at the west side of the depot and have since been removed.

John Tincher: Remembered there was a building in the woods where OD Green barrels with "U.S. Army" stenciled on them were stored. Tincher also remembered that other National Guard members were not allowed near the building, which he believed was numbered 517 and had signs and symbols of radiation on it.

MAP OF SCHOOL DISTRICT CONTAMINATION

The areas of toxic contamination identified by the Ohio EPA interviews have been plotted on a River Valley Local School District site map, designated with corresponding numbers across the campus. Number 11 represents the most volatile toxic chemicals that were located in and around the athletic field areas.

1. Batteries dumped in the landfill where the baseball diamonds were located.
2. Drums dumped into a ditch that ran the length of the depot toward the area where the schools are located.
3. 105mm Howitzer cannons were fired on what now was the school property.
4. Railroad ties, trash, trees, lumber and other materials were burned in an area near the schools.
5. Gasoline, fuel oil, antifreeze and motor oil drained and dumped in a pit and burned in the area of the schools.
6. Gas was dumped and burned in trenches where the high school is now located.
7. Most of the disposal took place in a large area on the east side of the depot where the schools are located.
8. Solvent waste drained into grate with tiles toward the east south of the school.
9. Sandblasting the Bailey bridges where the schools are now located and where radioactive reflective markers were left behind.
10. Arsenic waste found in the ditches.
11. Volatile organic compounds (VOCs) and TAL metals buried on the athletic fields.

The toxic waste site map illustrates what the employees of the former Marion Engineer Depot remember about the daily dumping of waste on what became the school district's middle school, high school and athletic field sites.

Interestingly, an article in the August 2001 edition of the *Family Circle* captures how the U.S. Army Corps of Engineers were trenching behind the schools and discovered an oily sheen spread across the dirt and a heavy gasoline smell a few feet below the soil. According to Jeff Steers of the Ohio

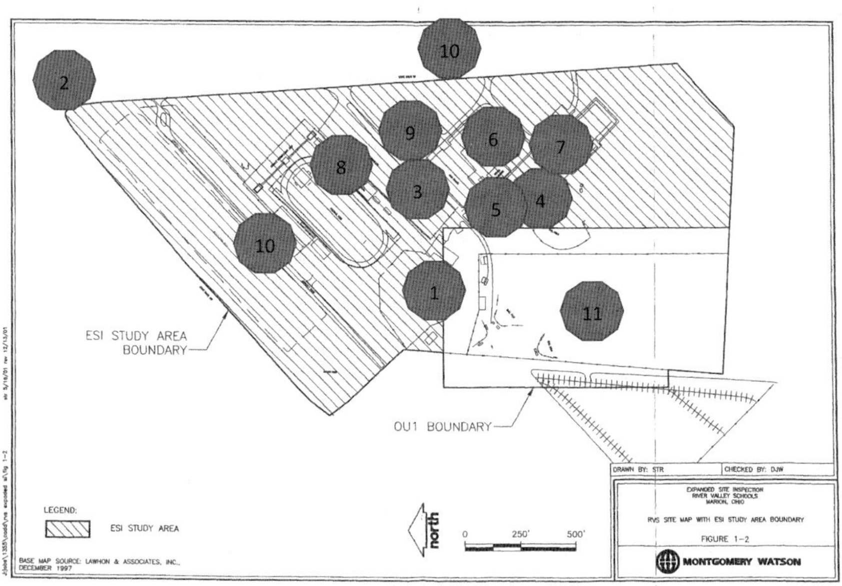

Site map the areas of toxic contamination identified by the interviews with the Ohio EPA officials. *Courtesy of the author's interpretation.*

EPA, this toxic mixture registered extremely high on their field instruments. Steers also discovered at the trenching site high levels of TCE and vinyl chloride, known human carcinogens that evaporate rapidly into the air. To compound matters, one of the investigator's radiation badges registered a high amount of radiation.[97]

OBSERVATIONS

In recent conversations and email correspondence with Gary Keller, a former superintendent of schools and friend of the author, Keller related that his father worked at the depot and recalled powerful solvents being used to clean the refurbished equipment. After cleaning the solvents, the liquids were dumped on the ground and leached into the soil. Keller's father also told him that they were not permitted to discuss what went on at the depot.[98]

The *Family Circle* article reaffirms what the interviewees stated about the dumping and burning of gasoline, fuel oil, antifreeze and motor oil in the area of the schools. Could the sandblasting of the Bailey bridges, with their radioactive reflective markers, have resulted in some of the markers being left behind and buried where the schools and athletic fields were built?

A more interesting question is: Did the school building contractors find toxic waste during construction of the schools and athletic fields, and if so, did they ignore it? Don Patterson believes that sandblasting and painting the Bailey bridges at the school sites may have resulted in contamination, with the debris being brought onto the school grounds while the schools were being built. Such contamination suggests that the building contractors either ignored or were unaware that the soil was possibly a mixture of toxic waste.

7
THE RECIPIENTS

I remember there was always a wet spot on the back of the field we ran through with wet socks and shoes.

The victims affected by the contamination of the River Valley Local School District tell stories of hardship and pain that they endured because of the pervasiveness of toxic chemicals at the school and athletic field sites, including a portion of the athletic fields that were built on top of a toxic waste dump, dating back to when the former Marion Engineer Depot was in operation from 1942 to 1961.

In 1997, it became apparent that the number of former River Valley Local School District graduates who contracted cancer seemed extremely high, given the size of the school district and the fact that the number of students in each graduating class was no more than two hundred. Jami Cummings's mother credits the nightmare their family went through as the reason for her involvement with the Concerned River Valley Families (CRVF). So too does Don Millard, Stephanie Millard's father, who became an activist in CRVF and later co-chaired the Restoration Advisory Board. The testimony of the cancer victims, who wondered whether the contaminated grounds of the River Valley Local School District contributed to their cancer, provides a narrative that possibly links the toxic grounds to their hardships.

Jami Cummings: A River Valley High School graduate, Cummings was diagnosed with leukemia at the age of twenty-three, which led to her eventually receiving a bone marrow transplant.[99]

STEPHANIE MILLARD: A 1991 River Valley High School graduate, Millard was diagnosed with breast cancer in 1997 at the age of twenty-four. Her father, Don Millard, remarked: "Cancer isn't supposed to happen to young people."[100]

KIM TOLNAR: A River Valley High School graduate, Tolnar was diagnosed in 1993 with chronic myelogenous leukemia and underwent a bone marrow transplant. She became involved in CRVF, which found through its studies that Marion had a 122 percent increase in leukemia deaths between 1966 and 1995, compared to just an 8 percent rise in Ohio over that time. Tolnar indicated that they "gained support of some residents but incurred the wrath of others who did not want to jeopardize Marion's reputation."[101]

AMANDA LOVETT: In the fall of 1997, Lovett, a sophomore, returned to school with her leukemia in remission, wondering if the school caused her disease but still wanting to attend classes.[102]

BETTY OWEN: Owen encountered hardships with breast and colon cancer, in addition to losing her forty-four-year-old son to brain cancer and her husband to lung cancer. At age fifteen, her son worked at the Central Machinery, which was built on the contaminated grounds of the former Marion Engineer Depot.[103]

ROBERT ANDERSON: Anderson, who worked as a fireman and guard at the former Marion Engineer Depot, contracted prostate cancer. He said, "They were always digging out there, but I never paid attention. I had no idea of the contamination on the land and I only knew there were some buildings they didn't want you to go into." The Anderson family built a home next to the contaminated U.S. Army Reserve property and indicated that if they knew of the contamination, they probably would not have bought the property.[104] Thus, for the Anderson household, the odds of contracting cancer were compounded by both Anderson's employment at the former Marion Engineer Depot and the proximity of their home to the dumping grounds of the reserve training facility.

MELANIE SERPA: A River Valley High School graduate, Serpa was diagnosed with Hodgkin's disease in 1985 while attending Miami University. She was cancer free for seven years, but when she was six months' pregnant, the cancer reoccurred with a mass growing in her shoulder. The doctors decided to induce labor, and her son was born six and a half weeks premature. She underwent chemotherapy ten days later.[105] Serpa wondered if the River Valley High School grounds contributed to her illness, given that she was on the cross-country team and the cheerleading squad, which put her into contact with the often-soggy school grounds. She remarked, "I remember there was always a wet spot on the back of the field we ran through with wet socks and shoes."[106]

TAMARA KEPLAR: Keplar, who worked for six years across the street from the former Marion Engineer Depot, was diagnosed with leukemia when she was eight months' pregnant and lost her baby after the first round of chemotherapy. Vince Keplar, her husband, said: "She's a forty-year casualty of the war."[107]

PATRICK PATTON: Patrick, a RVHS graduate, developed a brain tumor in 2005, first diagnosed as anaplastic astrocytoma and later as glioblastoma. He spent time on the athletic field areas, where he was in marching band and on the track team. The one memory about his track experience was that "we thought it was funny having a discus land halfway into the dirt then pulling it out, perfect landing, easy to measure the distance."[108] Patrick Patton's experience on the athletic fields mirrors Melanie Serpa's memory of often running on a soggy field.

PRISONERS OF WAR

A daily labor report of April 5, 1945, depicts the prisoner of war labor detail at the former Marion Engineer Depot.[109] It is interesting to see that the prisoners of war were assigned to such details as company and post drainage, paint shop, coal pile, painters, salvage yard and washing and lubrication of vehicles. These activities correlate to what we have learned about the contamination of the depot grounds, which later became a portion of the school district's middle school, high school and athletic field sites.

TABLE 5: MARION ENGINEER DEPOT DAILY LABOR REPORT

POWs	Type of Work Performed
12	**PW Camp** Maintenance on compound and company area drainage.
45	**Post Engineers** Post drainage detail, maintenance, labor on post railroad, handling, warehouse stock, blacksmith shop, paint shop and coal pile.

POWs	Type of Work Performed
119	**Warehouse Operations** General police of roundhouse, stacking dunnage, assembling cable racks, upkeep of depot stock, window washing, unloading equipment, painters, salvage yard labor, processing belts, assembling cables, stacking lumber clean-up crew and loading equipment.
30	**Maintenance Branch** Maintenance and repair of heavy equipment, washing and lubrication of vehicles.

OBSERVATIONS

The River Valley High School graduates are the unfortunate ones who suffered from possible exposure to carcinogens and other cancer-causing chemicals that contaminated the school sites, especially the athletic fields. What about the prisoners of war who underwent arsenic delousing and worked with toxic debris—did they end up developing cancer and related diseases once they returned to Germany?

In a telephone interview with Patrick Patton, Patton related to the author that his father focused on hiring a lawyer to examine the connection of the contaminated school sites and athletic fields to his son's brain tumor. What his father learned was that initially there was interest in pursuing a lawsuit, but once the parties discovered they would be suing the government, everyone declined to pursue legal action.[110] Unfortunately, for the recipients, the government agencies were deemed too powerful for lawyers to pursue in confronting them about contamination of the school district sites and trying to establish a causation to the cancer incurred by Patrick Patton and his fellow River Valley High School colleagues.

8

GOVERNMENT COVER-UP OR INCOMPETENCE

We have always said if nothing else, people can learn from Marion how not to run an environmental investigation.

A precursor to the environmental site investigations of the River Valley Local School District was what happened with the water supply at Camp Lejeune, North Carolina, from August 1, 1953, to December 31, 1987, where there was a high level of trichloroethylene (TCE), tetrachloroethylene (PCE), benzene and vinyl chloride found. Some parties believed that there was stonewalling by government agencies, including the U.S. Marine Corps, since the first discovery of contaminated wells was in 1980 and the wells were not shuttered until 1985.[111] As will be seen in this chapter, the delay in and slowness of the Camp Lejeune investigations was, tragically, a phenomenon that was repeated in the River Valley Local School District environmental site investigations, which began in 1997. It was not until 2003 that students were finally moved to new facilities. In total, it took approximately fourteen years from the first environmental site investigations to the completion of the cleanup of the former Marion Engineer Depot in 2012.

HINT OF CONTAMINATION

From what can be discerned, there was some hint of contamination problems at the former Marion Engineer Depot as far back as 1990, when an environmental site assessment of the depot grounds by ERM-Midwest Inc. indicated that there could be environmental issues, including underground storage tanks, waste, PCB transformer storage and the presence of radioactive materials. The assessment was done in response to a pending transfer of property to the Graham Investment/GP Prosperities. The conclusions by ERM-Midwest were based on past practices at the depot. The Environmental Assessment Report found the following:

- 114 buildings had stockpiles of tin, chromium, tannin, asbestos and radioactive materials;
- Surface water samples on the west side of the property where a brass/foundry was located showed findings of arsenic, cadmium, lead and toluene elements above detection levels.
- Building 106 had two tubes of radium 226 that were removed.
- Aerial photographs dating to 1951 showed an active disposal yard near a reservoir actively in use.
- Reviewed reports about asbestos removal and disposal of drums of waste of twenty-five fifty-five-gallon drums.
- On November 12, 1990, twenty-three smaller containers with composite samples of trichloroethane were moved.
- On December 11, 1990, thirty-seven pounds of transformers containing less than 450 parts per million (ppm) of PCB were moved.
- Other removals included 1,200 pounds of transformers containing greater than 450 ppm of PCB and 2,000 pounds of capacitors.[112]

The Environmental Assessment Report should have set off red flags that the school district property was possibly contaminated, but it was not until 1997 that suspicions arose about potential problems of contamination on the school district grounds when a county nurse found that there was a higher-than-normal rate of leukemia among the River Valley graduates.[113]

Pursuant to the ERM-Midwest report, the United States government took responsibility for removing the underground storage tanks, while GP Properties was responsible for removing the storage waste, PCB transformers, and radioactive materials before the property was transferred.[114]

ELEVATED CONCERNS

Concerns heightened throughout the River Valley Local School District community as the Ohio Department Health and the Ohio EPA started to conduct environmental site investigations into the possibility of contamination on the school sites. On February 2, 1999, both agencies declared that there were no health risks on the school grounds,[115] but at the same time, the U.S. Army Corps of Engineers started additional air monitoring at the site, which raised concerns about the possible premature assurances on the part of the state agencies. This confusion was compounded when a February 15, 1999 headline in the *Marion Star* read "Fallout from Toxic Barrels Continues." The article revealed that near the school grounds there were leaking toxic barrels of chromium and trichloroethylene, substances that were used as solvents for removing grease from metal parts. Polychlorinated biphenyl (PCBs), a known cancer-causing chemical, was found on the U.S. Army Reserve property.[116] PCBs were a coolant fluid used in electrical apparatus and most likely in the generators at the depot. Knowledge of the toxic barrels had been withheld by the U.S. Army Reserve for almost three months. The lack of notification about the chemicals was followed up by a letter to the parents from school superintendent Tom Shade, which read in part:

> *The grounds in question are approximately one-half mile away from the schools. The grounds are fenced off and the area where the barrels were found is 500–800 yards beyond the fence boundary. Given that, we are still concerned about the way the information was released about the Army Reserve grounds. We at River Valley were not told about the investigators' findings before last week, and we should have been. You should have been informed as well. This investigation depends not only gathering of valid data but also on the release of that data as soon as possible. Anything less is unacceptable.*[117]

Another article in the February 16, 1999 edition of the *Marion Star* read, "Marion deserves better from the Army Corps," which mirrored Shade's concerns that "the Corps should have come forward with the latest news about cancer-causing chemicals found at the Army Reserve training grounds. Why when it was finally released did it show up in the Marion Public Library files?"[118]

During this time, a media outlet called *Speak Ohio* was established as a site for posting about the River Valley Local School District contamination issues. Some of the postings on *Speak Ohio* reaffirmed how some people in the community felt about the handling of the toxic waste on the school district grounds. A parent posted that the U.S. Army Reservists' property was contaminated all the while that students were still attending school and participating in outdoor activities. The parent goes on to say: "It is so disgusting and obviously WRONG that if I had not witnessed it myself, I would find it hard to believe."[119]

Another posting reads: "We have always said if nothing else, people can learn from Marion how not to run an environmental investigation."[120] This posting once again highlights the frustration felt by the River Valley Local School District parents regarding how the U.S. Army Corps of Engineers and other government agencies were handling the potential contamination of the school district grounds.

On February 19, 1999, a headline in *Columbus Dispatch* read, "New EPA Director Takes Charge of Marion Probe: The Multiagency Investigation Has Produced Reports of Contamination that Were Not Released to the Public." The article goes on to list the following:

- A report documenting chemical contamination hundreds of times above acceptable levels just beyond the school campus was quietly slipped into the repository at the Marion County Public Library. The corps had prepared the report about the land owned by the U.S. Army Reserve.
- The EPA failed to tell the public that aerial photographs indicate that the River Valley Middle School may have been built on a dump.
- In March 1998, the Corps did not disclose that cancer-causing chemicals were detected during a dig on the athletic fields.[121]

These kinds of missteps led to anger and distrust on the part of the River Valley Local School District parents, who perceived this lack of communication as a possible cover-up on the part of the government agencies. The author believes that this was a prime example of bureaucratic incompetence that placed the students at risk. Another article, "Toxic Shock—Deceptive Public Servants Shirk Duties," in the February 21, 1999 edition of the *Columbus Dispatch* states: "The betrayal of the public trust in Marion is inexcusable."[122] This reinforces the idea of incompetence in the pursuit of finding answers about the role the toxic contamination played in the high rate of cancer among the students and graduates. A sense of incompetency and mistrust in the investigation were expressed in the March 23, 1999, the *Marion Star* headline: "U.S. Army Reserves Knew about Toxic Barrels 3 Years Ago."[123]

REJECTION OF FINDINGS

As discussed in chapter 2, Jed Ball raised concerns that the radiation testing results were doctored so that they appeared normal, but his claims were rejected by government agencies. Eventually, independent consultants were hired; they did not substantiate his claims, which lends credence to the notion that independent thinking was frowned upon by the government agencies.

An environmental toxicologist, Dr. Bruce Molholt, was hired to do an investigation and believed that the students and staff should be relocated based on the health risks. His report outlined the following:

1. Inadequate testing procedures had failed to detect high levels of vinyl-chloride vapors.
2. The estimated 3,600 gallons of chemical trichloroethylene used to clean metal parts were still present on a portion of the River Valley campus.
3. The presence of TCE and its breakdown chemicals are an environmental impact that must be addressed during the remedial effort.
4. A conclusion that there is a potential health risk and the students should be moved immediately.[124]

The Ohio Department of Health and Ohio EPA officials maintained that Dr. Molholt mixed cancer data with environmental data and therefore rejected the consultant's report. However, Dr. Molholt believed that there were gaps in the testing data collected by the U.S. Army Corps of Engineers and other government agencies investigating the contamination of the River Valley Local School District grounds, findings that, of course, were refuted by those agencies and the U.S. Army Corps of Engineers.[125]

It is interesting to see that the school district's environmental consulting firm also criticized the Corps' work and wrote a letter dated December 6, 1999, to the school district stating that "the Corps' reports misrepresent the actual test results, and in virtually every case, underreported the true extent of contamination."

A *Speak Ohio* posting references an ABC news report of how the U.S. Army Corps of Engineers had ordered an expert to change figures on a report so that it would benefit them to show that their project was not a waste of time.[126] A reporter responded to the posting by saying, "It is time the people of this nation stand up and once and for all, put a halt to the dastardly deeds of the U.S. Military. It appears they like, the IRS, answer to no one and can get away with whatever they so desire. The River Valley saga is a perfect example of just that. They, the military, couldn't have picked a better community to play 'their games', in the city in denial."[127]

COVER-UP OR JUST CAUTION?

In January 2002, a report was released stating that the students were not at risk because the indoor and outdoor air quality of the school district was consistent with EPA requirements. A U.S. Army Corps of Engineers interoffice communication in June 2001 raised some issues with the monitoring of the air quality of the River Valley Local School District, where a fairly high rate of methyl ethyl ketone (MEC) was found. MEC, a colorless liquid used as an organic solvent, can irritate the nose and throat and harm the nervous system if inhaled.

The text of the U.S. Army Corps of Engineers' communique is copied below and should be analyzed as to whether it signals merely caution or a deeper cover-up:

> *Preliminary Air Monitoring results have come in from our May 27 and May 28 ambient air monitoring event that includes Methyl Ethyl Ketone (MEK) at levels outside the instrument calibration range. This is a first such reading in the several air monitoring events that we had RVS. I have asked the PDT to identify potential MEK sources. My initial thoughts are that it would not be from Operable Unit (OU1). We will be challenged to convince the vocal few that will be concerned. I still need to verify that we have not detected MEK in the soils or groundwater, if we had any detections, they were limited. There are industries within the range of RVS that could be a source.*
>
> *At this point the data is preliminary and we should not comment outside our organization. Unless the holiday impacts us. I would expect that sometime*

> *next week I will start the notification process to Ohio EPA and RVS. Once the data goes to the Ohio EPA, RVS and the RAB we can expect some public activity. The timing of our release of this information and the Family Circle article (will) make this an interesting detection. I would expect the vocal few to once again utilize this leverage for immediate relocation.*[128]

In the minutes of the May 23, 2002 Restoration Advisory Board (RAB), J.R. Kolmer Engineering, consultant to the RAB, was critical of the investigation work plan for the former Marion Engineer Depot and Scioto Ordnance Plant. The firm's critical comments were (1) the work plan lacked detail about past investigations and site locations; (2) there was little to no discussion, except in the objective statement, and the RAB did not clearly state whether or not it was looking to verify contamination or whether there was need for an additional investigation; (3) previous results had been clearly stated in the work plan; and (4) some details were missing and no rationale was given for types, location and analytical methods of samplings being taken.[129]

In a *Marion Online* update about the River Valley Environmental Investigation, a resident thanked J.R. Kolmer Engineering for illustrating that the army was using faulty and inadequate data. The firm disagreed with the army's "composite" or "averaging" testing procedures, which was Jed Ball's premise from the beginning.[130]

WHISTLEBLOWER

In the early stages of the River Valley Local School District cancer scare, an Ohio EPA employee, Paul Jayco, was assigned to oversee the environmental site investigation. As he began his work, Jayco became concerned about contamination issues at the high school and middle school sites. He began sending memos to his superiors suggesting that air, soil and water tests be performed before the schools opened in the fall of 1998. The tests had been continually delayed until they finally began just before school started, at which point it was found there was contamination on the athletic fields, an area that was roped-off and later replaced by a chain-link fence. His other memos expressed concern over the high rates of cancer among River Valley graduates and the lack of response by government agencies in pursuing expedited investigations of the school district and to develop a plan of action for remediating suspected contaminants.[131]

After Jayco's memos were obtained by the press, the Ohio EPA employee received a ten-day suspension without pay, supposedly for drinking two beers at dinner before a meeting but in reality it was believed the suspension stemmed from retaliation for the negative attention the memos had brought on the Ohio EPA, including an investigation done by NBC's television program *Dateline*.[132]

An administrative law judge for the Labor Department in Cincinnati reinstated Jayco with back pay and full benefits and ordered the Ohio EPA to pay his legal fees. The judge stated: "Mr. Jayco cannot be punished for writing critical internal memos of an investigation of possible links between contaminated soil on the River Valley High School campus in Marion County and the number of leukemia cases among graduates."[133] It should be noted that Jayco was the first Ohio EPA employee to utilize whistleblower laws that were established to allow employees to bring forth safety-related issues.

Jayco's frustration with the slow pace of the Ohio EPA's action toward potential contamination issues at the River Valley Local School District speaks to the issue of whether the agency, like other government agencies, was either unprepared or incompetent in failing to take appropriate steps to safeguard the students in the school district. Mike Griffith, a geologist and environmental activist, states: "Paul took his job to serve and protect the citizens of the state seriously, while the bosses seemed to be more intent on protecting their image rather than protecting the children."[134]

Lack of Interagency Coordination

The General Accounting Office (GAO) reported that one of the major problems in the cleanup of Former Used Defense Sites was that the U.S. Army Corps of Engineers often did not coordinate with other government agencies, such as the EPA, during the cleanup phase and especially during the preliminary assessment. Surveys done by the GAO in 2003 show the lack of conectivity between state regulators and the U.S. Corps of Engineers.[135]

The GAO released its report in 2003, the same year that the River Valley Local School District was leaving its contaminated grounds and moving into new facilities with no known contamination and that were considered environmentally safe for the students to attend. The GAO's survey evidences a lack of communication between state and federal agencies during the preliminary assessment phase of Formerly Used Defense Sites cleanup projects, a communication breakdown that was similarly experienced in the River Valley Local School District case and the inconsistency of information and updates from state and federal agencies between 1997 and 2003.

TABLE 6: RIVER VALLEY MIDDLE AND HIGH SCHOOL IN RELATION TO THE FORMER MARION ENGINEER DEPOT

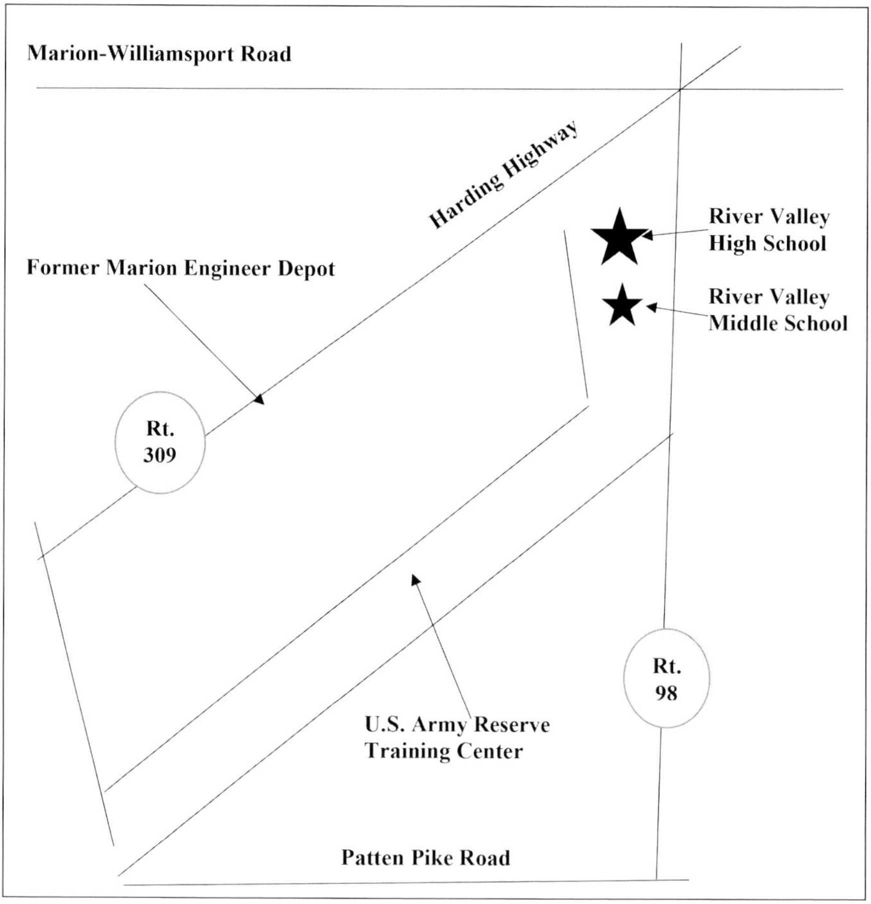

TABLE 7: STATE PROJECT MANAGERS' REPONSES REGARDING CORPS COORDINATION WITH STATES DURING PRELIMINARY ASSESSMENTS OF ELIGIBILITY IN THE SAMPLE

	YES	NO	DON'T KNOW	TOTAL RESPONSES
Did the Corps inform states that it was starting a preliminary assessment of eligibility?	6%	81%	13%	444
Did the Corps ask states for information or input on its approach?	6%	80%	14%	886
Did the Corps ask for state participation?	5%	84%	11%	441
Did the Corps inform states of interim results as worked progressed?	5%	83%	12%	441
Did the Corps provide states with a draft of the report summarizing the results of the preliminary assessment of eligibility?	4%	84%	12%	442
Did the Corps provide states with the final report of the preliminary assessment of eligibility?	48%	43%	9%	442

Additional GAO reports also show a continued lack of information sharing between state and federal agencies during Formerly Used Defense Sites cleanup projects. A 2010 GAO report found that the Department of Defense (DOD) often failed to disclose some contamination to the EPA and the public in a timely fashion. The report also found that the DOD opposed a recommendation that Congress consider amending Section 120 of the Comprehensive Environmental Response, Compensation, and Liability Act (CERCLA) to authorize the EPA to be the sole agency to administratively impose penalties to enforce cleanup requirements at federal facilities.[136]

A 2012 GAO report found the following additional shortcomings in communication: (1) limitations in the DOD's environmental restoration policies that may hinder its ability to identify and respond to exposures; (2) the DOD not fully documenting responses to public health recommendations; (3)

the DOD restoration program not specifying when to conduct public health assessments beyond the initial assessment at the National Priorities List sites; (4) the DOD environmental restoration policies not addressing the health risks from past exposures; (5) difficulty in documenting an environmental exposure; (6) difficulty in scientifically linking environmental exposure to health conditions; and (7) difficulty in seeking a remedy for an environmental exposure under the Federal Tort Claims Act.[137] A January 2017 GAO report further found that the DOD did not report to Congress in its annual report "contaminants that have a reasonable pathway to enter the environment, present a potential unacceptable risk, and do not have regulatory standards based on peer-reviewed science," which would preclude Congress from having cost information for cleanup of the affected sites.[138]

The bureaucratic posturing that the school district had to endure and what other Formerly Used Defense Sites projects have encountered may have led to more health-related problems for those associated with the projects because of the bureaucratic fumbling and infighting.

Was the infighting and positioning by state and federal agencies more incompetence than cover-up? Let us consider what the graduates of the River Valley Local School District who are cancer survivors and those with health-related problems would say.

OBSERVATIONS

It was tragically unfortunate that the Ohio Department of Health, Ohio EPA and U.S. Army Corps of Engineers did not always communicate their findings in a timely fashion to the River Valley Local School District parents and graduates. At times, government agencies would tell the community there were no health risks to the students, only to soon reverse course and announce that they found contamination on the school grounds. Despite first roping off the athletic fields and then installing a chain-link fence, the officials continued to reassure the community not to worry and that it was safe for the students to attend school and after-school events, all the while periodic tests showed evidence of contamination on the school grounds.

The GAO reports illustrate a lack of coordination between state and federal agencies and infighting and positioning that still exists, most recently referenced in a 2017 GAO report of Formerly Used Defense Sites projects.

Frustration with the way the investigations were proceeding by government agencies was reflected at the September 26, 2001 RAB meeting when the board questioned the lateness of the Ohio Department Health's reporting of leukemia among River Valley High School graduates and in Marion County. The meeting minutes expressed this concern: "It appears the longer they waited to release the more they revised to downplay the statistically high leukemia cases from people connected to the school and Marion Engineer Depot."[139]

One has to remember that calls to investigate the environmental conditions at the site started in 1997 when the graduates and parents of the school district began to push the government agencies to do an in-depth study of the school district grounds. This movement lasted until 2003, when the new middle school and high school buildings were ready to be occupied. Credit also needs to be given to Paul Jayco of the EPA, who continued to press his employer to take seriously the contamination issues on the River Valley middle school, high school and athletic field sites.

A major question remains: During the six years of deliberations and environmental site inspections, were personnel and students exposed to health risks that were linked to cancer and other diseases? In retrospect, the authorities should have heeded Dr. Bruce Molholt's recommendation to relocate the students to another location, such as the Tri-Rivers Joint Vocational School, and to force the government agencies to pay for the temporary relocation until the new schools were built. Based on his prior experience as a school superintendent, the author recognizes that relocating students during a construction project is no easy task, but it can be accomplished.

In the end, the plight of the River Valley Local School District site contamination seems to originate more from incompetence and lack of coordination rather than a cover-up by the government agencies.

9
"GIVE ME A BREAK"

My disappointment is they say nine kids who graduated from the high school (had leukemia) and then they have the nerve to say they can't find a link. Give me a break. The only thing I can surmise is the fear of litigation. Truth is there, if people would just look for it.

The claim by government agencies that there were not any chemicals found on the River Valley Local School District grounds linked to leukemia and cancer of the graduates is questionable. An investigative report in August 1997 by the Floyd Brown & Associates Environmental Company of thirty-eight soil samples, including five groundwater and one drinking water sample, found that there was no significant contamination. A subsequent headline in the August 26, 1997 edition of the *Marion Star* read "OPEA River Valley Is Safe," which sent a message to the public that there was no cause for concern and that it was a safe environment for students to attend school.[140] Yet soon after, in October 1997, the Ohio Department of Health stated that "environmental factors seem to be to blame for the higher-than-average leukemia rates in the county." Dr. John Wilson, assistant director at the James Cancer Center Hospital and Research Institute at The Ohio State University, provided more encouragement to parents despite the lack of current direct evidence: "Researchers have to study trends over several years to get answers to reasons behind the leukemia rates, but conclusive evidence can be reached."[141]

The problem the investigators faced was determining that what existed at the time the air, soil and water sampling was performed does not necessarily represent the chemical exposure paths ten or twenty years previous, although it did lead to the conclusion that there was no direct link between chemical exposure to high rates of leukemia.

The chemical exposure pathway to humans is dependent on five factors: (1) source, (2) environment transport, (3) point of exposure, (4) route of exposure and (5) people who might be exposed.[142] Key to the students' problems was determining whether they encountered chemicals in the soil or air and whether what occurred (route of exposure) was through drinking, eating, breathing or touching the chemicals, especially those students who participated in activities on the soggy athletic fields.

An article in the October 12, 1998 edition of the *Columbus Dispatch* held that finding an environmental cause of cancer was almost impossible to prove and went further in saying that scientists estimated only 8 percent of cancers came from environmental exposures while 65 percent of cancers were caused by smoking or poor eating habits.[143]

ENVIRONMENTAL TRANSPORT

In 1961, when the River Valley Local School District purchased seventy-eight acres from the former Marion Engineer Depot, there were no federal environmental regulations in place, so the U.S. Army was responsible for overseeing matters related to any possible contamination at the site purchase by the school district. Because the military installations were so integral to the war effort, there was likely little concern over the environmental impact, and so, too, were environmental issues just gaining a nascent awareness in the United States.

Up until 1992, with the passage of the Federal Facility Compliance Act (3), U.S. military installations were exempt from complying with U.S. environmental regulations. The past practice of burn pits and waste dumps containing oil, gasoline, greases, solvents, PCBs, arsenic, heavy metals and other carcinogens that were either emitted or discharged into the soil, air and water, continued on and were migrated to the acreage purchased by the school district.

Dr. Ted H. Schettler, in his paper "Reverberations of Militarism: Toxic Contamination, the Environment and Health," states:

> *Soil contaminants may remain in the surface, adhere tenaciously to soil particles at varying depths, or leach downward, finally reaching groundwater*

> *that, in turn, migrates over time. Soil contaminants may also be taken up by plants through the roots, entering the consumable portion of a crop. The complex mixture of contaminants found on many military and industrial sites is therefore, dramatically moving through the environment, presenting increasing and persistent opportunities for human exposure through direct skin contact, ingestion, or inhalation.*[144]

The pathways that existed at the former Marion Engineer Depot—dispersed chemicals via burn pits, toxic waste dumps, leaking barrels and contaminants mixed in the soil and open toxic waste trenches—over a period of time permeated the soil prior to the construction of the middle school, high school and athletic field sites, which may have contributed to long-term health problems resulting from human exposure to the contaminants.[145] It is reasonable to believe that the contamination of the former Marion Engineer Depot site from 1942 to 1961 left behind a high degree of contaminants in the soil and possibly in the air that may have adversely affected the health and welfare of the personnel and students of the River Valley Local School District. Essentially, the community of the River Valley Local School District did not stand a chance. The following diagram depicts the likely contamination exposure pathways that negatively affected the students and personnel of the school district between 1963 and 2003, representing some forty years of undue suffering because of unregulated chemical disposal and the low priority given by the government to environmental pollution. In fact, it was not until 1997 that the Concerned River Valley Families group pushed government agencies to investigate the school district grounds for contaminants left behind by the former Marion Engineer Depot, which was originally under the management of the Department of Defense (DOD) and the U.S. Army.

TABLE 8: CORPS PROJECT MANAGERS' REPONSES REGARDING CORPS COORDINATION WITH STATES DURING PRELIMINARY ASSESSMENTS OF ELIGIBILITY IN THE SAMPLE

	YES	**NO**	**DON'T KNOW**	**TOTAL RESPONSES**
Did the Corps inform states that it was starting a preliminary assessment of eligibility?	24%	58%	18%	481

	YES	NO	DON'T KNOW	TOTAL RESPONSES
Did the Corps ask states for information or input on its approach?	27%	53%	20%	965
Did the Corps ask for state participation?	16%	64%	20%	481
Did the Corps inform states of interim results as worked progressed?	15%	65%	20%	478
Did the Corps provide states with a draft of the report summarizing the results of the preliminary assessment of eligibility?	7%	79%	14%	481
Did the Corps provide states with the final report of the preliminary assessment of eligibility?	56%	30%	14%	477

TABLE 9: REPRODUCTIVE DISORDERS ASSOCIATED WITH ENVIRONMENTAL EXPOSURE

Lead	Stillbirth preterm delivery, low birth weight, spontaneous abortion, mental retardation and spermatotoxicity.
Mercury	Spontaneous abortion, blindness, deafness, mental retardation, delayed development and brain damage.
Benzene	Spontaneous abortion and low birth weight.
Chlorinated solvents (TCE, PCE, DCE)	Eye, ear, oral cleft and cardiac abnormalities.
Polychlorinated biphenyls (PCBs)	Preterm delivery, low birth weight, reduced head circumference and growth deficiencies.

TABLE 10: HEMATOLOGICAL AND IMMUNE SYSTEM ALTERATIONS ASSOCIATED WITH ENVIRONMENTAL EXPOSURE

Arsenic	Leukopenia and eosinophilia..
Beryllium	Granulomatous hypersensitivity.
Asbestos	Depressed cell-mediated immunity, increased auto-antibodies including anti-nuclear antibodies.
Mercury	Immune complexes deposited in the kidney.
Benzene	Lymphocytopenia, autoimmunity and aplastic anemia.
PCBs	Depressed immunoglobulins, atrophy of Lymphoid tissue, decreased antibody response, increased susceptibility to infection.
Solvents	T-cell abnormalities and autoimmunity.

TABLE 11: ADVERSE NEUROLOGICAL EFFECTS ASSOCIATED WITH ENVIRONMENTAL EXPOSURES

Arsenic	Peripheral neuropathy.
Lead	Encephalopathy (high levels), peripheral neuropathy and neurobehavioral impairment.
Mercury	Central and peripheral neuropathy.
Solvents (TCE, PCE)	Subtle neurophysiological and neuropsychological effects at low levels.

Given the exposure pathway, a latency period always exists before disease detection, and it appears that the River Valley Local School District's suspected leukemia cluster matches the Latent Chart.[146] A lengthy exposure exists from point A to C1, which matches the period in which the students were participating in activities on the contaminated athletic fields on the school grounds, and the external exposure ended once the students graduated from high school. Then, a true latent period is established as illustrated from point B to D, with the presumptive period marking the manifestation of the suspected leukemia cluster and possibly other diseases over a longer period versus a per-year basis.

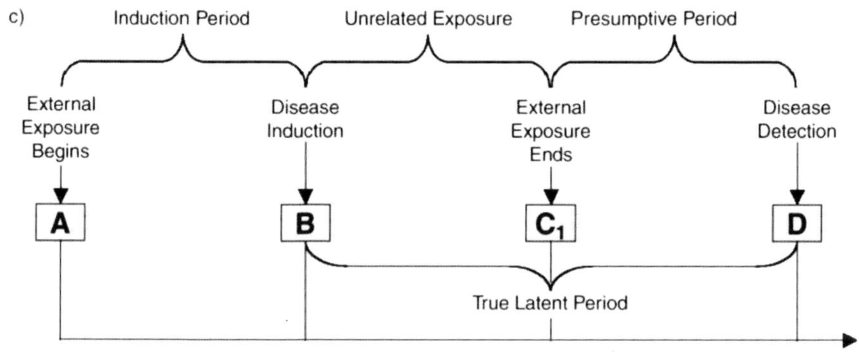

Evaluation of latent and presumptive periods. *Courtesy of the NCBI, 2004.*

The site map developed in chapter 6 (see page 68) illustrates an interpretation of the areas of toxic contamination on the school grounds based on the recollections of former Marion Engineer Depot employees, painting a vivid picture of the expansiveness of the potential contamination on what later became the middle school, high school and athletic field sites.

The interviews point to the locations on the school sites and the former Marion Engineer Depot site where waste dumps and burn pits contained batteries, gasoline, oil, antifreeze, motor oil, solvent waste, arsenic, semi-volatile compounds (SVOCs), volatile organic compounds (VOCs) and TAL metals, plus the identified carcinogens with corresponding toxic substance rankings in chapter 4, see Table 2: Carcinogenic to Humans Found on the School Grounds with Toxicity Rankings (page 49).

OBSERVATIONS

My observation can be summarized in the history of a middle school, high school and athletic fields that were inundated with toxic chemicals that may have had a negative effect on the health and welfare of the students who attended classes in those two buildings and who participated in extracurriculars on athletic field sites built on a toxic waste dump.[147] Ultimately, the environmental site investigation reports were written with such technical jargon and in a format that in most cases were too detailed to

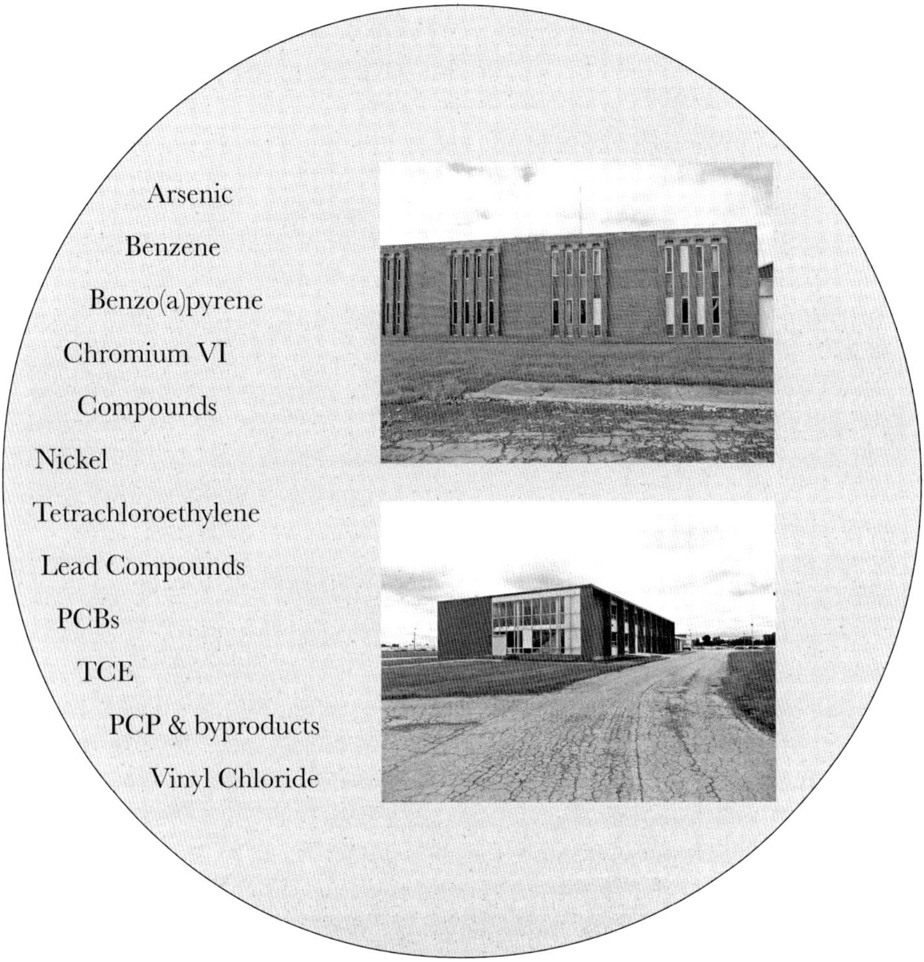

Depiction of the middle school (*top*) and high school (*bottom*) surrounded by toxic chemicals. *Courtesy of the author's collection and interpretation.*

understand by the average person. The scientific language used resonated with a sense of information control and power that put the authorities in control and instilled a false sense of safety to the community. The section on Environmental Transport clearly shows what environmental exposure pathways of over forty years can do to the human body because of unregulated chemical disposal and the nonexistence of environmental pollution regulations. My research involving the River Valley School Local District and the community fits with Dr. Wilson's belief that "conclusive evidence can be reached."[148] Kim Tolnar, a 1983 RVHS graduate diagnosed with leukemia ten years later, summarizes this broken state of affairs: "My disappointment is they say nine kids who graduated from the high school (had leukemia) and then they have the nerve to say they can't find a link. Give me a break. The only thing I can surmise is the fear of litigation. The truth is there, if people would just look for it."[149]

10
ADDITIONAL EVIDENCE

Students were issued water bottles at the start of the 2001 school orientation and were instructed to fill the bottles at home and bring them back to school.

The environmental site investigation reports indicated that the potable water piped from the Ohio American Water Company wells to the middle school and high school was safe, but hidden evidence reveals the possibility that the fifty-eight-year-old water lines may have leaked because of deteriorated joints, which would have allowed contaminants to enter the water supply.

An April 1, 1999 article in the *Columbus Dispatch* related that the Ohio American Water Company shut down two of its sixteen wells after tests revealed unsafe levels of hazardous chemicals in the wells, triggered by the Ohio EPA water-well testing as a part of the environmental safety investigation of the River Valley Local School District.[150]

In a *Speak Ohio* posting on June 19, 2000, it was revealed that a new water line was installed from the east end to the west of the former Marion Engineer Depot, raising the question of whether the line was connected to the schools and whether the disturbed soil was tested for contaminants. If the new water line passing through toxic soil would break, would there be the same situation that might have occurred had it been a fifty-eight-year-old water line connected to the schools?[151]

Paul Jayco, an Ohio EPA consultant and eventual whistleblower who was assigned to coordinate the environmental site inspections of the school district, sent out warning signals about the potential for contaminated water going to the schools. Jayco questioned the thoroughness of the Ohio American Water Company's testing parameters, which did not address numerous cancer-causing agents. He believed that the company's testing parameters were limited and should have included a full metals, pesticides and semi-volatile analysis to include total polycyclic aromatic hydrocarbons (PAHs) produced from coal, crude oil and gasoline. Jayco's other concern was the raw water intake from the Little Scioto River, located a quarter of a mile from a contaminated area of the river. A Biological, Sediment and Water Quality Study performed on April 4, 1994, found that there were seventeen PAHs, of which five were known or suspected carcinogens, in addition to twenty-three metals and cyanide measuring in at highly elevated levels. His warnings fit with what the Restoration Advisory Board Technical Subcommittee learned about the possibility of joint leaks in the fifty-eight-year-old water lines servicing the schools.[152]

At the September 17, 2001 Restoration Advisory Board Technical Subcommittee meeting, one of the main discussion points was the fifty-eight-year-old drinking water line going through the OU1 (athletic fields) contaminated waste area to the schools. According to the U.S. Army Corps of Engineers' data, it was believed that the water line might be leaking. The subcommittee questioned whether a potentially contaminated water line was safe for students to use. It was brought to the attention of the subcommittee that trenching and digging were prohibited in the OU1 area because of exposure to the waste disposal area. Water supplies at the school sites were provided by the Ohio American Water Company public water supply system, with no groundwater resources at the site being used by the school district.[153]

The subcommittee was also informed that water lines are not always pressurized. A lack of pressure in the lines causes the joint seals to be compromised and deteriorate in the presence of solvents, a scenario that would allow contaminants to enter the water lines during low pressure periods. The subcommittee was given piping drawings of the water line going through OU1. It was believed that an electromagnetic survey verified that the water line going through the OU1 waste disposal area to the schools showed disturbances that might be caused by the water line leaking and provided the opportunity for contaminants to enter the lines.[154]

It was also indicated to the subcommittee that trenching was started by the U.S. Army Corps of Engineers. During its investigation, the corps encountered solvents with escaping volatile chemicals, which forced the corps to discontinue its trenching. The subcommittee questioned why the drinking water line that ran through or near the waste disposal area continued to be used.[155]

The subcommittee learned that students were issued water bottles at the beginning of the 2001 school year during orientation and were instructed to fill the bottles at home and bring them to school each day, but data shows that the bottle usage was low and most students disregarded the request.[156]

Given the electromagnetic survey and the students being asked to bring filled water bottles to school, it was recommended that a new water line be installed from Route 98 to the schools. The existing fire hydrant that was fed from the former depot could remain and be used for only fire service. The other recommendation was that a new line should be used for potable drinking water, showers, dish washing and the like, and the water line should be approximately four hundred feet long from Route 98.[157]

As previously indicated, the OU1 site was an area used for the disposal and burning of debris, paints, fuels, oils and solvents from the former Marion Engineer Depot. Carcinogens found at the disposal site were trichloroethene (TCE) and tetrachloroethene (PCE), the same volatile organic compounds found in the drinking water at Camp Lejeune. All told, service members there were exposed to contaminants for thirty-four years, while River Valley Local School District students may have been exposed for up to forty years, all from a water line fifty-eight years old with deteriorated waterline joints.

At Camp Lejeune, there was evidence that linked contaminants found in the water supply to eight certain diseases. Some veterans were diagnosed with conditions that were linked to their military time at the camp, which led to a disability compensation system. Can the same be said for the River Valley Local School District, whose students may have been exposed to the same compounds that may have contributed to a suspected leukemia cluster and other cancers and could thus be categorized as presumptive diseases?

In the various government reports documenting the River Valley Local School District contamination, there does not seem to be any mention of water lines connected to the school that ran through the OU1 waste dump area and the potential for contaminated drinking water due to leaking joints. Was there a potential for contaminated drinking water, and could this have been another contributor to the high leukemia and cancer rates of the River Valley Local High School graduates? Why weren't any red flags raised at the start of the 2001 school year when all students were given water bottles and asked to fill them up at home and bring them to school, despite the majority of the students not following the request?

OBSERVATIONS

On page 49, Table 2: Carcinogenic to Humans Found on the School Grounds with Toxicity Rankings lists the carcinogens arsenic, benzene, benzo[a]pyrene, beryllium, chromium, ionizing radiation, lead compounds, pentachlorophenol (PCP), polychlorinated biphenyls (PCBs) trichloroethylene and vinyl chloride, all of which are a part of the ninety contaminants identified as harmful to drinking water in public water systems based on the Safe Drinking Water Act of 1974, as amended in 1986 and 1996 and regulated by the U.S. EPA.[158]

The question remains: Prior to the building of the schools and athletic fields, did the wells on the Marion Engineer Depot grounds leak carcinogens, the likes of which were identified in Table 2, into the groundwater? Furthermore, did those contaminants permeate the middle school, high school and athletic field sites, leaving the students at risk while attending school from 1963 to 2003? Paul Jayco asserts that the thoroughness of testing procedures failed to address numerous cancer-causing agents, and the Restoration Advisory Board subcommittee learned that there was a possibility for joint leaks at low pressure periods in the fifty-eight-year-old water lines servicing the schools. Thus, it raises the question whether the public water system was out of compliance with the Safe Drinking Water Act, subsequently leaving the students open to ingesting cancer-causing agents.

Did the River Valley Local School District contamination also include compromised water lines like what happened at Camp Lejeune? Did anyone listen to whistleblower Paul Jayco's warnings about the potential for contaminated waterlines to the schools because of limited testing parameters by the American Water Supply Company? Did PCBs, PCPs, TCEs and other solvents seep into the water lines, sending contaminated potable water to the middle school and high school? Why wasn't there more frequent testing other than once-a-year random samplings of the drinkable water at the schools, and why wasn't the sampling done at low-pressure periods in order to get a more accurate measure if any contaminants may have been entering the water lines during those periods of time? What about the drinking water fountains near the concession stands that showed elevated levels of lead and trihalomethanes? Were the water lines to the schools out of compliance with the Safe Drinking Water Act?

There are enough unanswered questions that could lead one to believe that the potable water supply to the schools was at times contaminated because of inadequate testing parameters, leaking water line joints and raw water being drawn from the highly contaminated Little Scioto River. This possibility of water line contamination could be one additional contributing factor to the development of a suspected leukemia cluster and other cancers contracted by the graduates of River Valley High School.

11

SUSPECTED LEUKEMIA (DISEASE) CLUSTER

These students have a right to know they went to school for the last six years on top of a toxic waste dump.

DISEASE CLUSTER

A disease cluster is defined as an occurrence of a greater than expected number of cases of a disease among members of a specific group, residents of a specific area or over a specific period. The elements of a disease definition are as follows: (1) cluster means an excess of disease, (2) greater than expected number of cases means that there is a statistically significant excess of cases when compared to an established baseline and (3) particular disease means an outbreak of a specific form of cancer within a community.[159]

In June 1997, the Ohio Department of Health (ODH) investigated a report of a high number of leukemia cases among River Valley High School (RVHS) graduates and students. This high number of people afflicted with leukemia is classified as a disease cluster when it is located in a certain place and time, which in the case of the River Valley Local School District was thought to be caused by potential ground contamination at the middle school, high school and athletic field sites.

SUSPECTED LEUKEMIA CLUSTER

The River Valley Local School District suspected leukemia cluster was similar to the Woburn, Massachusetts leukemia cluster that was discovered in May 1979, when several wells—the principal source of water for East Woburn—were found to have large concentrations of trichloroethylene (TCE) and tetrachloroethylene (PCE), organic compounds known to cause cancer in laboratory animals.[160]

The EPA investigated the Woburn site and found dangerous levels of lead, arsenic and chromium in a local wetlands area but did not immediately tell the local government or public. The *Woburn Daily Times* found out about the toxic chemicals and broke the story in September 1979. Given the slow response time of the government to react to the perceived "cancer cluster," a local citizens group called For A Cleaner Environment (FACE) placed ads in the newspaper seeking a response from those in the community who had childhood leukemia. The responses were compiled and yielded twenty-eight identified leukemia cases between 1964 and 1989. The Massachusetts Department of Public Health (Massachusetts DPH) identified eighteen leukemia cases among all age groups between 1968 and 1979.[161] Even though the Massachusetts DPH essentially refuted the data from FACE, the data collected by the group mobilized the Woburn residents to become more proactive to find out the cause of the high rate of leukemia. They suspected the high cancer rate was a result of contaminated ground water and toxic chemicals found in the local wetlands. The community efforts eventually yielded results, including a settlement reached in 1986 by chemical producer W.R. Grace with five families over health problems other than leukemia, followed by a $69.5 million fine by the EPA in 1991 against Beatrice Foods, W.R. Grace, Unifirst, Hemingway Trucking and New England Plastics to clean up two well sites.[162] If FACE had not been relentless in pursuing these industrial giants, the corporations would have walked away without being held accountable for contaminating the city wells and wetlands of Woburn, Massachusetts.

Back in Caledonia, Ohio, the Concerned River Valley Families (CRVF) was equally as persistent as the Woburn community in searching for the cause of a suspected leukemia cluster, perhaps from the contamination of the school district grounds. The River Valley parents, like the Woburn parents, pursued the causes of the suspected leukemia cluster and developed a distrust of the government agencies for failing to be forthcoming with the public about their findings.

The Ohio Department of Health's study in July and August 1997 showed that there were six observed cases of leukemia, which they said "revealed a statistically higher number of confirmed leukemia cases than expected based on national background rates." Don Millard, the concerned parent who was behind the ads ran by the CRVF, identified the following results, evidencing that there were more than just leukemia cases among the graduates:

- seventeen breast cancers
- twelve leukemias
- eleven melanomas
- five brain cancers
- four ovarian cancers
- two skin cancers
- two colon cancers
- three cases of Hodgkin's disease[163]

After Don Millard released his self-reported results, he stated: "The point of placing the ad, and attempting to gather information of River Valley graduates is to prove agencies like ODH that there are substantial number of cancer cases, not just leukemia associated with the schools." Millard went on to say, "These students have a right to know they went to school for the last six years on top of a toxic waste dump."[164]

The CRVF believed that there was some kind of environmental stress affecting the schools that could be due to the contamination left behind by the former Marion Engineer Depot and the Scioto Ordnance Plant. Proactive measures were undertaken by the CRVF, which included running ads out of a concern that there may be some unreported cases.[165] Interestingly, in a 1997 national survey of state health department requests for identifying cancer clusters by state, Ohio was the only state that did not participate in the survey.[166] Consider whether this absence by the Ohio Department of Health was a precursor to the River Valley Local School District environmental site investigations, whereby the agency had essentially predetermined that there were no cancer clusters in the state and no need for a study that wouldn't find evidence of contaminants or carcinogens on the school grounds that would have caused the suspected leukemia cluster and other cancers afflicting RVHS graduates.

FREQUENCY OF LEUKEMIA DEATHS

An article in the October 17, 1997 edition of the *Columbus Dispatch* proclaimed: "Leukemia news stuns Marion—Death rates up 122% in 30 years, state study finds." In comparison, the death rate from leukemia in Ohio over the same period was only 8 percent. Keep in mind the Marion County population count does not include the population of Marion City.[167] The spike in leukemia in Marion City from 1984 to 1995 may have resulted from environmental exposure that occurred years before the deaths.[168]

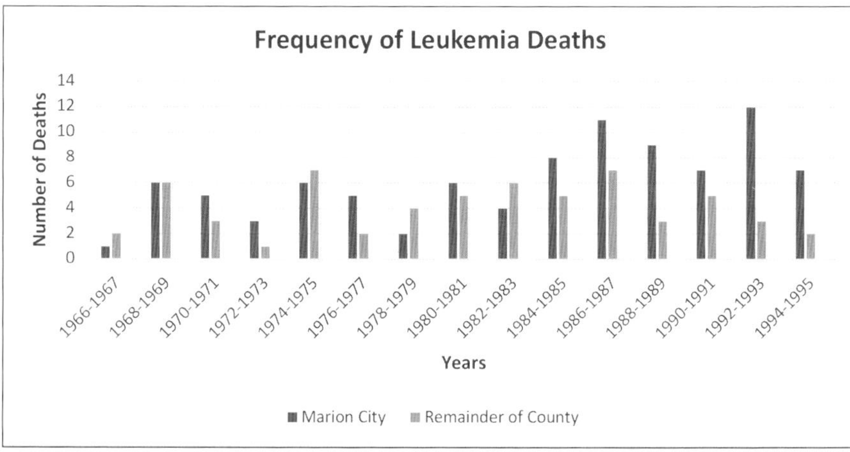

OHIO DEPARTMENT OF HEALTH STUDY OF 2001

The incidence of leukemia among RVHS graduates shows that the types of contacts with the depot grounds occurred not only through school attendance but also through contact with the athletic fields that hosted football, baseball, track, cheerleading and sports statistician activities (six cases, 67 percent), while the Future Farmers of America participation resulted in at least two cases, or 22 percent of the total.[169]

The major reason for the ODH study was to identify whether the graduates of RVHS were at an increased risk for leukemia as compared to the general population of Marion County. If the students at the high school were found to have been at an increased risk of leukemia, then the study was to assist in identifying potential risks factors at the high school. Unfortunately, because of the limitations of the study imposed by the Ohio

Table 12: River Valley Local School District Contamination Exposure Pathways

Sources
Burn pits, toxic waste dumps, leaking barrels, contaminates mixed in the soil and open toxic trenches.

↓

Environmental Avenues
Soil, sediment, groundwater, air, plants, and surface water.

↓

Exposure Point
78 acres purchased from the former Marion Engineer Depot.

↓

Exposure Route
Contamination entering the body through eating, drinking, breathing and touching toxic substances.

↓

Receivers of the Contamination
Students and personnel absorb toxicants that target specific organs leading to clinical dseases.

Department of Health, those risk factors were never identified, presumably because it was determined to be too difficult to identify potential cancer-causing agents that may have played a role in the development of leukemia among RVHS graduates.

Dr. Bruce Molholt's Feedback

Dr. Bruce Molholt, a subcontractor, was hired by the Restoration Advisory Board to review the following ODH study: "Case Study of Leukemia Among Residents of Marion County, Ohio, 1992–1999, and Graduates of River Valley High School, 1963–2000 (ODH, 2001)."[170] Dr. Molholt agreed with the ODH that there was no evidence the leukemia victims in the study had more exposure, either directly or indirectly, to tobacco smoke because it was his belief that the River Valley patients tended to be younger than leukemia patients elsewhere in the country and thus were much less likely to have smoked.[171]

Dr. Molholt further agreed that the soil contact at the high school has been a parameter to the increased risk for leukemia but disagrees with the ODH's approach to include Marion County residents who worked at the former Marion Engineer Depot because it was his belief that by doing so would distort the percentage of high school graduates who contracted leukemia and other cancers. He disagreed that further study of leukemia among high school graduates and the Marion County residents would not be useful, and he believed that a further study was needed that would include separate data concerning RVHS graduates versus nongraduates.[172]

In a previous ODH study, only 32 percent of 5,345 graduates who matriculated between 1963 to 1999 responded. It was Dr. Molholt's belief that the remaining 68 percent, or 3,619 graduates, of nonresponding RVHS graduates should have been pursued so as to add to the 2001 study. Dr. Molholt questioned why the more than two-thirds of nonresponding graduates were not followed up with; in failing to do so, it limited the pool of graduates for the 2001 study.[173]

According to Dr. Molholt, a limiting factor in the ODH study was that the Ohio registry began on January 1, 1992, thus limiting the measurement of the prevalence of leukemia in Marion County from before that time. Also, for this review, the registry was only up to date through December 31, 1999, further limiting the leukemia data for Marion County by encompassing only eight years out of the total thirty-seven years of graduates.[174]

Dr. Molholt believes that the flaw in the design of ODH's study undermines the ability to distinguish risk factors in RVHS graduates versus nongraduates. He stated:

> *Age at leukemia diagnosis is the most striking statistical difference between Marion County resident and RVHS graduate groups. 20–39 age bracket is enormously increased among RVHS graduates. 3/42 (7%) cases in Marion County residents vs. 6/9 (67%) in RVHS graduates. This disparity may be even greater since 4 RVHS graduates were included in the Marion County residents.*[175]

Dr. Molholt acknowledges that there is a latency period between exposure to carcinogens and a cancer diagnosis. Given the high proportion of graduates who have developed leukemia between the ages of twenty to thirty-nine— 67 percent versus 7 percent in Marion County overall—this marked a tenfold increase in the identified age group. Thus, Dr. Molholt concludes that RVHS students were more intensely exposed to carcinogens than other members of the community who did not attend the high school.[176] It should be noted, too, that students at the middle school were not allowed to go out the back doors and thus did not have direct access to the contaminated grounds, which further speaks to the concerns expressed by Dr. Molholt that students were being exposed to potential carcinogens at the school sites.

LEUKEMIA STUDY CESSATION QUESTIONS

The cessation of the study leaves open additional questions about why the study did not include three students who attended RVHS but did not graduate; why the study did not reference the fact that the average age of graduates studied was twenty-nine, versus an average age of fifty-eight for the general Marion County community, a disparity that could very well have undercounted the incidence of leukemia among the graduates had they developed the disease later in life; and why didn't the study expand its pool of those studied beyond the eight graduates and one student identified in an earlier survey by going back and resurveying the 68 percent of graduates who did not respond to the survey? In the end, the study disregarded the evidence of potential links between exposure to contamination at the school site and leukemia.

LEUKEMIA PRIMER

In response to the cancer concerns raised by the residents of Marion County, the Central Chapter of the American Cancer Society released a leukemia primer as a means of educating the River Valley community about the four kinds of leukemia and provide general data to assist the community in understanding the different kinds of leukemia that they had seen in the graduates of the school district.[177] It is interesting to see that the causes in the table include both radiation and chemical exposure, which could have been the leading causes cancer in the River Valley Local School District.

Case Review of Leukemias Diagnosed Among Residents of Marion County, Ohio, 1992-1999 and Among Graduates of River Valley High School (RVHS), Marion, Ohio, 1963-2000, by History of Exposure to the Marion Engineer Depot/Scioto Ordnance Plant[1]

	Marion County Residents[2]		RVHS Graduates	
	Cases	Percent	Cases	Percent
Total Cases	42	100%	9	100%
Contact with Depot Grounds				
Total Cases Reporting Depot Contact	14	33%	9	100%
Cases Reporting Substantial Contact[3]	6	14%	9	100%
River Valley Graduate	4	10%	9	100%
Athletic Fields	5	12%	6	67%
Future Farmers of America Activities	0	0%	2	22%
Worked at Depot/On Depot Grounds				
Employment at Depot	0	0%	0	0%
Worked on Grounds[4]	4	10%	0	0%

[1]Source: Community Health Assessments Section, Ohio Department of Health, May 2001.
[2]Includes the four RVHS graduates who were Marion County residents diagnosed with leukemia from 1992-1999.
[3]Cases may be represented in more than one subgroup.
[4]Types of work include: Telephone pole/line installation for GTE; deliveries to grounds for GTE; work for DeKalb Genetics Inc.; and substitute teaching in district for two days.

MEETING DISEASE CLUSTER CRITERIA

When one reviews the totality of the data, it seems likely that there was a leukemia cluster and at a minimum a suspected leukemia cluster in the River Valley Local School District caused by the contamination of the school grounds, especially from the carcinogens found in and around the athletic fields.

A cancer cluster "is a greater-than-expected number of cancer cases that occurs within a group of people in geographic area over period."[178] To be considered a suspected leukemia cluster in the River Valley Local School

District, the data must meet the following criteria[179] of the Centers for Disease Control and Prevention:

- *A greater than expected number*: The River Valley Local School District had a greater number of leukemia cases as would be observed in a similar setting, evidenced by death rates that were up 122 percent over a thirty-year period compared to the overall death rate for leukemia in Ohio of just 8 percent over the same period.[180] An Ohio Department of Health study found that in 1997 there were six cases in Marion County, which they said "revealed a statistically higher number of confirmed leukemia cases than expected based on national background rates."[181] Another verifying data point in this chapter found that six, or 67 percent, of RVHS graduates studied who had leukemia had contact with the athletic fields for football, baseball, track, cheerleading and sports statistician, while two, or 22 percent, of those surveyed had participation in the Future Farmers of America.[182]
- *Of cancer cases*: For this criterion, the cases of cancer must involve the same type of cancer and have the same cause of the disease, which in the case of the River Valley Local School District was leukemia caused by exposure on the athletic fields and school grounds to vinyl chloride, benzene, benzo[a] pyrene, trichloroethylene and other carcinogens. According to the Agency for the Toxic Substances and Disease Registry Substances Priority List of 2017, vinyl chloride is ranked the fourth most toxic chemical, benzene sixth, benzo[a]pyrene eighth and trichloroethylene (TCE) sixteenth.[183]

An internal EPA memo revealed that the athletic fields had carcinogenic contaminants one thousand times higher than acceptable levels, including polycyclic aromatic hydrocarbons (PAHs) (chemicals that occur as a byproduct from coal, crude oil and gasoline), benzo[a]pyrene (chemicals that come from burned substances such as oil and gasoline) and vinyl chloride (used to make polyvinyl chloride (PVC) plastics). The memo also found trichloroethylene (metal degreasing solvent banned in some states) some four thousand times higher than average detected about a foot below the subsurface soil near the girls' softball field. Furthermore, volatile organic compounds (such as benzene) were discovered on the athletic fields; they perhaps came from the buried burn pits and waste dumps of the former Marion Engineer Depot.[184]

- *That occurs within a group of people*: The graduates were Caucasian who had grown up in Marion County and were in their twenties and thirties when their leukemia occurred, which matched the race, ethnicity and age of the population type for Marion County.
- *In a geographic area*: This criterion was met with nine cases of leukemia between 1963 and 2000, with six graduates who participated on the contaminated athletic fields and two graduates who participated in Future Farmers of America activities, which likely took place on the contaminated school district grounds. The geographic boundaries where the graduates were exposed to carcinogens that led to their leukemia was certainly defined to a confined area, namely, the athletic fields and middle and high school grounds.
- *Over a period of time*: The period of time was 1963 to 2003, which fits the cancer cluster definition of identifying the period of time for the evolution of the cancer cluster or suspected leukemia cluster.

OBSERVATIONS

Mr. Millard's and Dr. Molholt's questioning of the Ohio Department of Health data that failed to find a direct link to the leukemia cluster mirrors the same concerns that the Woburn parents held toward the Massachusetts government agencies for seeming to stonewall and manipulate the data. In the case of the suspected leukemia cluster in the River Valley Local School District, it was perhaps purely bureaucratic incompetence and cost avoidance to find out the real truth. Most concerning were the doubts raised by Dr. Molholt that the final study may have been flawed since it did not consider missing data and appeared to display an unwillingness to do a more expansive survey collection over a longer period of time, which raises the question of subjectivity and the decision or lack thereof to initiate or conclude a cluster study.

The data comparisons of leukemia among the River Valley graduates certainly raises speculation that there was environmental stress in the River

Valley Local School District. The full data analysis seems to at a minimum meet the CDC's criteria for a cancer cluster and/or suspected leukemia cluster among the identified group of graduates of the school district who were from the same geographic area and had the same age range, race, ethnicity and similar population strata.

It seems that the outcome was possibly predetermined, given the fact that the Ohio Department of Health was the only state health department that did not participate in a 1997 national study of state health systems in identifying cancer clusters by state and thus sent the signal that it was the only state that did not consider the concept of cancer clusters.

It is quite possible that the environmental conditions of the school sites may have worsened by the time the schools were occupied in 1963, given the activity on the covered toxic waste dump that was in and around the athletic fields. So too was the soil disturbed while the schools were being built, which set up the perfect storm for an exposure pathway and direct link from the contaminants and carcinogens to the development of a cancer cluster and suspected leukemia cluster and other forms of cancer that afflicted the RVHS graduates. Given the latency periods for some of the cancers, the question remains whether the students who were attending the middle school and high school up until their closing in 2003 were subjected to the same debilitating diseases of the earlier RVHS graduates.

12
MORE QUESTIONS

Was there negligence with a breach of duty of care with the sale of the former Marion Engineer Depot property to the River Valley Local School District?

If one follows the data, a sequence of events emerges that shows the likely path of contamination of the River Valley Local School District grounds beginning in 1942 with the opening of the Marion Engineer Depot and continuing until its closing in 1961. During its years of operation, the depot used burn pits; toxic waste dumps containing oil, gasoline, chemicals and other debris; sandblasting of Bailey bridges that left paint chips on the grounds; and the delousing of German POWs and cleaning of railroad cars with arsenic as its byproduct. Don Patterson, who inspected many of the buildings on site, attested to this contamination. He remembered the Bailey bridges being stored where the schools were eventually located, recalled the bridges being sandblasted and repainted and believed the problem at the schools stemmed from the debris of the bridges being brought onto the school grounds while the schools were being built.[185]

Chapter 2, "A Trail of Radioactive Materials," traced how metascopes (or sniperscopes) and radium markers were stored in lead-lined buildings that at times suffered leaks and needed to be remediated, which calls into question whether radioactive materials were left behind on the depot grounds and possibly migrated to the school grounds and buried in the construction sites of the middle school, high school and athletic fields.

In chapter 4, Table 2: Carcinogenic to Humans Found on the School Grounds with Toxicity Rankings and Table: 3: Carcinogenic to Humans by Organs, outlines the types of carcinogens that may have negatively affected the middle school and high school students. An article in the *Columbus Dispatch* on May 31, 2017, "Questions Linger Decades after River Valley Cleanup,"[186] captures the doubts held by the residents as they continued to question the outcome of years of environmental site investigations that found no scientific link between the illnesses in the community and the toxic waste and carcinogens found on the school districts campus. The questions may have been answered if the Concerned River Valley Families had the resources to successfully sue the U.S. Army Corps of Engineers and the DOD for the contamination of the school grounds that would have linked the toxic exposure to the suspected leukemia cluster and other cancers of the graduates. Such an outcome would have disavowed the presumption that there was no link between the toxic waste and the debilitating diseases of the RVHS graduates.

TRANSFER OF PROPERTY

The question remains: did the U.S. Army and the DOD have a legal duty to provide the school district with a safe and habitable seventy-eight acres that was free from contaminants?

Another lingering question is whether the school district bears any responsibility for ensuring that the school grounds were free from contaminants at the time it purchased the property in the early 1960s. Under the legal doctrine of caveat emptor, or "let the buyer beware," the buyer bears the responsibility for ensuring that the property is in suitable condition. Unfortunately, in this case it appears that the school grounds were full of contaminants that were either undiscovered or ignored by all parties.

Given the interviews of the former Marion Engineer Depot employees, we now know that people were aware that the school grounds were once a dumping site of toxic waste. It thus begs the questions: did the River Valley Local School District Board of Education either miss or ignore the contamination? Was knowledge of the contamination ignored because the seventy-eight-acre site cost only $22,350 or $286.53 per acre?[187]

Today, when federal real property is transferred to a non–potentially responsible party, the deed must contain notice of the type and quantity of hazardous substances and the time at which the storage, release or disposal occurred. In addition, the parties must disclose if remedial actions or disposal of any hazardous waste took place. Before the transfer is completed, the federal government must issue a covenant that warrants that all remedial actions necessary to protect human health were done.

Given the property transfer requirements that were later implemented, would it have been wise for RVHS claimants to argue that the contamination data found on site was proof enough that the federal government failed to provide a safe and habitable school environment for the middle school and high school students? Certainly, what the pattern of data shows is that the students and staff were subjected to a barrage of various levels of contamination on the school district grounds. What is most disturbing are the numerous human carcinogens that were found on the property that are linked to different forms of cancer.

NEGLIGENCE

Was there negligence in the sale of the school district property in the early 1960s by the U.S. Army and DOD? Negligence requires a breach of duty of care, and a duty may arise when a party creates an unreasonable risk of harm to another.[188] Was there negligence to a breach of duty of care when the former Marion Engineer Depot property was sold to the River Valley Local School District?

In the case of the River Valley Local School District contamination, did the cancers, diseases and emotional stress among the afflicted graduates stem from over forty years of having the students attend schools on contaminated grounds? What about the six-year period between 1997 and 2003 when the students were still attending school in a known contaminated setting of toxic waste and carcinogens?

CAUSATION

Let's examine the case of the River Valley School District graduates being diagnosed with cancer by considering the questions that come into play as to a possible court proceeding based on the following:

1. *What kind of chemicals were the individuals exposed to?* The data and information laid out in this book make the case that there was contamination, including carcinogens, at various sites and levels throughout the school district grounds, most notably in the danger caused as the students ran through toxic sledge on the athletic fields.
2. *Was there expert testimony based on scientific methods that were established and accepted?* The experts Dr. Bruce Molholt and Jed Ball maintain that there was a contamination problem on the school grounds that may have negatively affected the students, but unfortunately their analysis was refuted by the federal government. Had their assessment been accepted, the case could have been made that there was likely a strong association between the cancer and other diseases and the contaminated school grounds.
3. *Given the similar characteristics of the students who lived in Marion County, is there another source that would have caused their diseases other than chemicals found on the school grounds?* Despite certain demographic and social factors such as cigarette smoking, when one considers the levels of contaminants found on the school grounds, a reasonable person would conclude that it was more than likely the contamination, above and beyond the social factors, that was the predominant cause of the cancers among the graduates of RVHS. Other supportive information came from the *Columbus Dispatch* article of October 17, 1997, titled, "Leukemia News Stuns Marion—Death Rates up 122% in 30 Years, State Study Finds," which compared the overall leukemia death rate of 8 percent in Ohio over the same time period. Furthermore, the Marion County statistic did not include the population of Marion City, where there was a spike in leukemia deaths between 1984 and 1995, which may have resulted from environmental exposure that occurred years before.[189]
4. *Was the students' exposure to the suspected chemicals more than likely the cause of the suspected leukemia and disease cluster?* Had Dr. Bruce Molholt's and Ted Ball's findings been accepted, it could have led to the conclusion that the graduates contracted their diseases from exposure to the identified carcinogens. Thus, a correlation would be drawn showing that exposure to the contamination increased the plaintiffs' disease risk, bolstering the argument that the RVHS graduates' diseases were more than likely caused by the toxic substances.

LIABILITY

The Comprehensive Environmental Response, Compensation and Liability Act (CERCLA) was created in 1980 to impose liability on parties responsible for, in whole or in part, the presence of hazardous substances at a site. The fund liability is triggered if the following occurs:

1. Hazardous wastes are present at a facility;
2. There is a release (or possibility of a release) of these hazardous substances;
3. Response cost have been or will be incurred; and
4. The defendant is a liable party.[190]

In the case of the River Valley Local School District and the former Marion Engineer Depot, all of the indicated liabilities were met, which meant the U.S. Army and U.S. Army Corps of Engineers were liable for the cleanup costs. As previously noted, ultimately $8 million was spent to clean up the school grounds once the schools were closed, $2 million to clean up the U.S. Army Reserve site, $15 million to assist with the relocation of the schools and $35 million to remediate the former Marion Engineer Depot site, which was then converted to an industrial park.

It is evident that there was liability on the part of the U.S. government for the cleanup of the school grounds and the former Marion Engineer Depot site, but there was never any link to the liability for the health problems of the graduates from exposure to carcinogens on the school grounds.

A study of the Camp Lejeune service members found that they were exposed to contaminated drinking water from August 1953 to December 1987, and it was further discovered that the contaminants included the volatile organic compounds trichloroethylene (TCE), perchloroethylene (PCE) and vinyl chloride, in addition to benzene. The U.S. Department of Veterans Affairs established a compensation program under what it called the Presumption Disease Rule that became effective March 14, 2017. Under the rule, the agency identified eight presumptive diseases and works under the presumption that specific conditions diagnosed in certain veterans were caused by their military service and that the circumstances of his or her service caused the condition.[191]

Imagine if the same rubric would have been applied to the River Valley Local School District graduates who were exposed to carcinogens on the school grounds that could have caused leukemia and other cancers. Such a determination could have marked a precursor to a presumptive disease compensation program for other Superfund and Formerly Used Defense Sites.

BARRIERS

The documentation of an environmental exposure is difficult to prove because the requirements often focus on the specific time, place and level of exposure. In the case of the River Valley Local School District, this linkage proves problematic since the time frame is over forty years. One major question that comes into play stems from the period between 1997 and 2003: were the toxic waste dumps on the athletic fields closely monitored? Furthermore, there was the installation of a chain-link fence, as well as a missed opportunity to collect data on the level of environmental contamination on the athletic fields, both of which could help trace the contamination to the graduates' health problems.

The other barrier to scientifically linking the contamination to cancer and other adverse health effects is based on the latency periods that make it difficult to establish the exact time of exposure, which may have been some years before the actual health problems became manifest. Once again, however, given the full breadth of the data, it seems logical that a reasonable person would conclude that contaminants found on the school grounds were indeed more than likely a major contributor to the various forms of cancers that afflicted the graduates of RVHS.

HEALTH RISK REIMBURSEMENT PROGRAM

Consider whether the DOD would realign its defense budget to provide remuneration to nonmilitary personnel who follow the Presumptive Disease Rule. Doing so would allow for compensation of civilian diseases that resulted from environmental exposure, which would then eliminate the cumbersome tort process, expense and time incurred by the afflicted individuals pursuing restitution.

The EPA's health risk assessment under the Superfund law is meant to identify the types of adverse health effects that are caused by exposure to some agent in question and to characterize the quality and weight of evidence supporting the identification.[192]

Another part of the EPA's health risk assessment is to identify the range of exposure.[193]

The EPA's range exposure chart illustrates those populations that have a high degree of contact with toxic substances over long periods of time

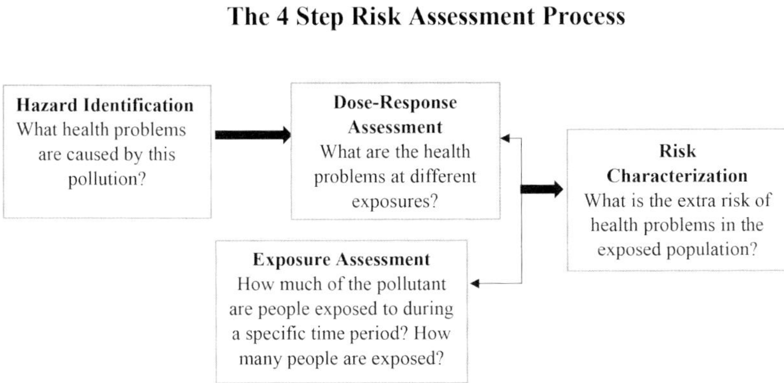

4-Step Risk Assessment Process. *Courtesy of the EPA, June 20, 2017.*

compared to populations with a lower degree of contact over shorter periods of time.

The problem with the health risk assessment process is that it is often hard to provide accurate exposure data. It is difficult to trace the effects of contaminants—plus, trying to use data from animal studies proves challenging in correlating the results between animal subjects and humans. Given that this process does not identify risks to a specific individual or group of individuals, the federal government is left to go through an extensive environmental health risk assessment for Superfund cleanup purposes. However, in this scenario, no actual causation of identified harm or exposure can be made, as in the case of the River Valley Local School District sites, to link the carcinogens found to an individual or group of individuals.

What if instead the federal government added another section to its risk assessment process that provides a revenue stream to those afflicted with debilitating diseases associated with exposure to high concentrations of carcinogens, specifically, those who qualify under the four-step assessment process and under the range of exposure that is found to be in the *area with ground level concentration 10 percent or more above the standard.* If a risk assessment identifies a Superfund or Formerly Used Defense Site, dollars for cleanup purposes would trigger the new section in the Superfund or Formerly Used Defense Sites programs that allocates if an individual or group of individuals hits the 10 percent threshold, they should be entitled to go through an initial screening process. Those individuals could ultimately be eligible for the same risk assessment review process that was allocated for the cleanup process. If they are found to meet all criteria under the

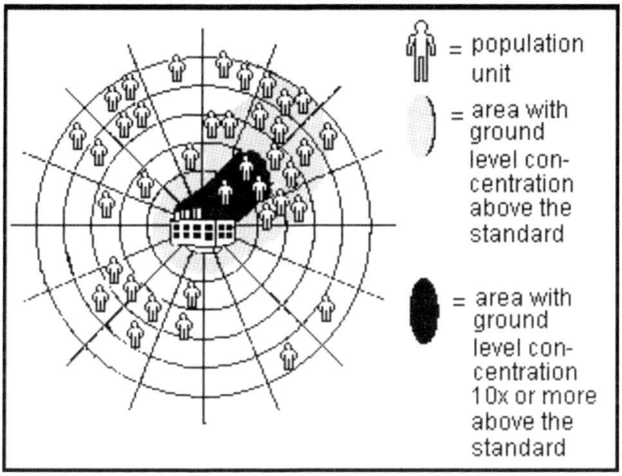

Range of exposure. *Courtesy of the EPA, June 20, 2017.*

scientific and medical evidence establishing an association between the chemical exposure, identified carcinogens at the site and development of specified diseases later in life, it would trigger a reimbursement plan similar to that in the Camp Lejeune program.

Once again, if this had been established for afflicted River Valley Local School District graduates, the reimbursement plan should show a causal link between the carcinogens and the debilitating disease, as those listed in the Carcinogenic to Humans by Organs Table 3 in chapter 4. Individuals would be eligible for healthcare costs related to one or more of the presumptive diseases, specifically, those cancers listed here that have been linked to the carcinogens found on the River Valley Local School District grounds:

Bladder	Colon	Liver	Skin
Blood Disease	Kidney	Lung	
Brain	Leukemia	Lymphoma	

Another consideration that should be integrated into the reimbursement program is the possibility a chemical release may cause an immune system disorder that triggers diseases not related to the chemical exposure.

The hypothetical health risk reimbursement program presented may come too late for the River Valley Local School District graduates, but it could be a start to a pilot presumptive disease reimbursement program for civilians subjected to debilitating diseases related to the contamination of Superfund and Formerly Used Defense sites.

OBSERVATIONS

It is quite possible that had the fact patterns outlined in this chapter been pursued in the form of a lawsuit, it would have provided an avenue for the afflicted graduates to receive some kind of restitution. Had the environmental site exposures been successfully linked to their corresponding diseases, it would have forced the DOD Environmental Restoration Program—which previously focused only on cleanup as a way to prevent future problems—to create a pathway toward assessing health problems from previous exposures and to identify those individuals who may have been afflicted from such exposures.

It seems evident that at the time of the River Valley Local School District environmental site investigations, there was no intention by the DOD's Environmental Restoration Program to provide restitution or guidance to afflicted individuals based on prior exposures. If that were the case, the DOD would have incurred not only the cleanup costs but also the restitution costs, which would have had major budgetary implications for the DOD.

One unfortunate piece of the River Valley story is that it is a community of average-income workers who did not have the financial resources to challenge the government assumptions through legal action. A protracted legal battle against the government would have indeed been challenging, but employing a reasonable person legal approach could well have been able to prove that there was a more than likely chance that the contamination on the school district grounds was one of the major contributors to the graduates' various debilitating diseases, including a suspected leukemia cluster.

13
YOU BE THE JUDGE

The findings were difficult to understand, and I hope that you are able to clarify the results.

The government's environmental site inspection reports contained scientific and technical jargon that the average person could not understand and clouded the real facts present. During the author's research in the Marion County Public Library, he encountered a resident who asked what he was researching. The author explained he was looking into the contamination of the River Valley Local School District, and the individual replied: "The findings were difficult to understand, and I hope that you are able to clarify the results."

Consider again the comment from chapter 2 where a former employee at the Marion Engineer Depot indicated there was radioactive waste dumped into the dirt and covered up where the schools were built. This comment is at the heart of the differing opinions about the possible contamination of the school grounds and the impact it had on the health and welfare of the graduates of RVHS. Thus, it leaves the reader to be the judge of the data and facts outlined in this book.

CAMP LEJEUNE REVISITED

By reflecting on the suspected disease and leukemia cluster and contamination of the River Valley Local School District grounds—especially the athletic

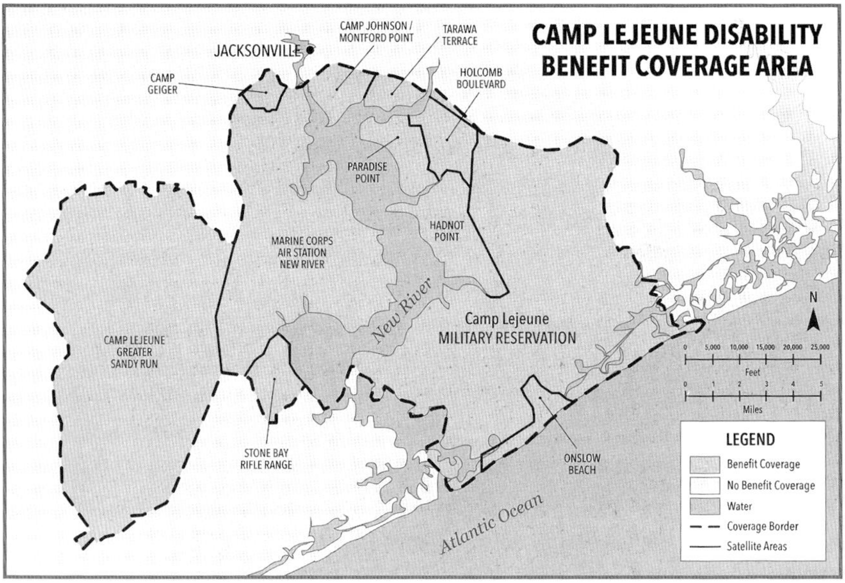

Camp Lejeune Disability Benefit Coverage Area. *Courtesy of the U.S. Department of Veterans Affairs, 2018.*

fields—that were left behind by the former Marion Engineer Depot, it places the RVHS graduates' plight into the same category as the contamination at Camp Lejeune in North Carolina. A diagram of the camp shows the compensation coverage areas for those marines who drank contaminated drinking water and is illustrative of the expansiveness of the contamination.[194]

Camp Lejeune can be compared to the River Valley Local School District toxic dilemma, whereby the school grounds were contaminated with carcinogens, especially the athletic fields that were being built in and around a toxic waste dump as evidenced by the U.S. Army Corps of Engineers. After the schools were closed and relocated in 2003, the government spent $8 million to clean up the school grounds. Like Camp Lejeune, was there a failure to detect the possibility of joint leaks in the fifty-eight-year-old water lines servicing the schools? Were the testing procedures limited by omitting a full metals, pesticides and other semi-water analysis to include a total polycyclic aromatic hydrocarbons test? Was there neglect in the River Valley drinking water testing procedures? You be the judge!

BURN PIT

Imagine what a burn pit or toxic waste dump would have looked like as a baseball field was being built on top of it, or school buildings being built on grounds that were once burn pits and toxic waste dumps.[195]

This stark depiction sends a clear picture of what the athletic fields may have looked like on top of a burn pit, with the outcome one of a persistent disease scenario at the school district sites. You be the judge!

Top: School baseball field. *Courtesy of the author's collection.*

Bottom: Tornado—Greensburg, Kansas Burn Pit. *Courtesy of the NARA, September 29, 2005.*

CONTINUOUS CONTAMINATION

From the beginning, questions were raised about the results of the environmental site investigations that failed to draw a straight line from the school district contamination to the suspected leukemia cluster and the number of cancer cases among the school district graduates. The government agencies believed that the science was incomplete and often referred to the fact that the incidence of cancer could have been caused by smoking or unhealthy diets.

The article "Toxic Troubles—What Toll Taken Public Health" in the October 4, 1998 edition of the *Columbus Dispatch* discusses the environmental site investigations of the River Valley Local School District, with several assertions made about the potential for bias in the investigation results. One of the assertions is that the "health department is reactive, not proactive in its approach to investing cancer hot spots" and another is that "unlike in Ohio, health officials in some other states keep a close eye on the cancer rates in environmentally troubled areas."[196] Recognizing that this article was published in October 1998, it appears that it may have foreshadowed the predetermined conclusions reached by the Ohio Department of Health, Ohio EPA and the U.S. Army Corps of Engineers, which all said there could be no established link between the school grounds contamination and the illnesses of the graduates.

Continuous contamination is seen in the recent Monsanto cases involving Roundup, whereby the chemical glyphosate is seen to be a direct contributor to the plaintiffs' cancer. The case has brought to light a revelation in 2015 that the International Agency for Research on Cancer had classified glyphosate as a probable carcinogen to humans, contrary to the conclusion reached by the U.S. EPA, with some suspecting that the agency's findings was in fact a cover-up tactic by a former employee.[197]

The question in the Monsanto manipulation case mirrors the River Valley Local School District story as to whether the government manipulated public opinion by disregarding the significant role that long-term exposure to carcinogens may have played in the development of cancer among the school district graduates. The problem of finding for restitution in these kinds of situations is determining whom to sue and whether it is worth the time and expense to sue the federal government for alleged cover-up or perceived neglect.

In another recent lawsuit specifying continuous contamination, it was found that the DOD had allowed a firefighting foam to slip into the drinking water systems for decades at military bases in the United States and abroad. This contamination resulted in the exposure of tens of thousands of people to a chemical called PFAS that is linked to cancers and other related health problems. After one lawyer advised that it would be too difficult to bring suit against the federal government, many plaintiffs opted to sue companies that made the product, such as 3M and Tyco Fire Products.[198] The same held true for the father of RVHS graduate Patrick Patton, who searched for lawyers to sue government agencies because he believed his son's brain tumor was caused by the school district's contaminated grounds, only to be told by the lawyers that it would be too difficult to fight the government in court.

The story of the River Valley Local School District is one in which students continued to attend school for six years after the first discovery of contamination in 1997. It is a story in which certain parties deferred to the experts who maintained that there were no serious health risks, despite the adamant contention of Dr. Bruce Molholt, Jed Ball and others who continued to believe that the health risks posed by the contamination on the school grounds was too high and possibly responsible for the suspected leukemia cluster and other cancers that afflicted the graduates. You be the judge!

PREVIOUS CHAPTERS' CONTAMINATION HISTORY

These previous chapters have provided a history of the contamination of the school grounds and illuminated a possible cover-up or mere incompetence on the part of government agencies as the environmental site investigations progressed. There is the additional question of whether the students could have been drinking contaminated water due to potential joint leaks and the lack of proper testing over a long period.

The large population of those who suffered leukemia and other forms of cancer leads one to believe that there was a greater than 50 percent chance of a disease cluster, in particular a suspected leukemia cluster, whereby there were more afflicted individuals compared to baseline data. Chapter 12, meanwhile, raises questions about the legal liability and restitution to the afflicted graduates.

It was the work of the Concerned River Valley Families who forced government agencies to develop better communication processes regarding the progress of the environmental site inspections in order to provide the public greater information about the site inspections.

The Restoration Advisory Board provided an accountability forum for the government agencies to report their environmental site investigation results and at times pushed for better timelines and understanding of what the government agencies were trying to say. One would hazard to guess that at times the board members believed they were getting a sales pitch to try to convince them that the report outcomes were right and not to dispute them since the government agencies had the final say, despite the repeated questioning of the findings by Dr. Bruce Molholt and Jed Ball, who believed that some of the governmental testing procedures were flawed.

An example of digging deeper behind the government viewpoint was the outspoken voice of Jim McGuire, chairmen of the technical subcommittee, who responded to an Ohio EPA press release by saying, "I am asking you to consider other people's suggestions. Do not represent yourself as the definitive authority. We have limited experts, limited instrumentation and limited technology. We only have to look back a few years to recognize how uninformed we were. I can only deduct that we are still uninformed."[199] McGuire's conclusion matches the issues raised in chapter 8, "Government Cover-Up or Incompetence," when it appeared at times that there was a lack of interagency coordination at the expense of the potential for students and personnel to be subjected to suspected disease or leukemia clusters. What was needed was immediate environmental site investigations that, unfortunately, may have been too late for some because of the six-year lag before the schools were relocated.

Another tragedy was the suspension of Ohio EPA site representative Paul Jayco, who vocally raised concerns about suspected contamination of the school grounds. Jayco needs to be credited with forcing government agencies to expedite environmental site inspections out of his fear that students and school personnel may be exposed to contaminated school grounds, including the athletic fields. He pushed for full air, soil and water tests to be performed before the opening of schools in fall 1998, which unfortunately did not happen. In a review of the record, Jayco was found to be a team player with excellent evaluations. In the case of the suspected contamination at the River Valley Local School District, however, it is clear that he became frustrated with the government's prolonged dragging of its feet and putting the students and school staff at risk of exposure to cancer-causing agents on the school grounds. You be the judge!

SUSPECTED DISEASE PYRAMID

The pyramid that follows illustrates in a straightforward way what happened in the River Valley Local School District over a forty-year period. The foundation of the pyramid represents the contaminated grounds of the former Marion Engineer Depot, leading further toward a suspected disease cluster and suspected leukemia cluster.

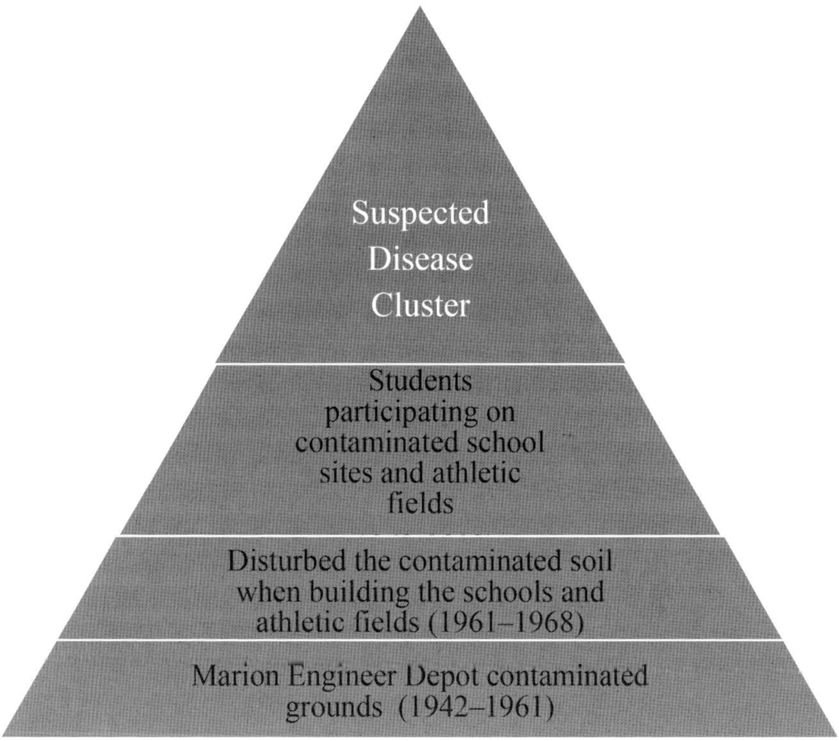

As stated earlier in the dedication to the graduates of the River Valley Local School District who battled cancer: "This is not a report clouded by scientific and technical jargon, but a presentation of understandable facts that contradicts previous conclusions." In other words, the seventy-eight acres of property of the former Marion Engineer Depot that was sold to the River Valley Local School District should have come with a detailed recordkeeping of the potential for contaminated acreage based on past practices at the depot from 1942 to 1961.

Some party is at fault for subjecting the students and school personnel to a suspected contaminated school environment for close to forty years that may have led to a suspected leukemia cluster and other debilitating diseases. The students were short-changed by the adults who neglected to take substantial action until Paul Jayco and the Concerned River Valley Families forced the government agencies to eventually relocate the schools to a safer, inhabitable learning environment. You be the judge!

IGNORING TRUE LATENT PERIODS

The adolescents who attended the middle school and high school were going through many physical changes, including growth spurts and organs and tissues that were still developing. This dynamic growth and development period put the adolescents at greater risk from environmental toxins, especially the carcinogens found on the school grounds. Under the true latent periods, the potential exposure to carcinogens by adolescents should have raised a red flag for the government agencies that were doing the environmental site investigations, since cancer can take decades to develop. It appears, however, that this latency period was ignored because the agencies continued to maintain that there was no established link between the contaminants found on the school grounds to the graduates' cancer cases. If the officials in charge of the environmental site investigations would have charted the true latent periods of the River Valley graduates' cancer cases, there is a strong likelihood that it would have resulted in a reasonable person concluding that the manifestation of the cancers was more than likely caused by the carcinogens and other contaminants found on the school grounds.

The work done by the government agencies was limited to the identifying of and eventual removal of contaminants from the school grounds without consideration of the negative impact on the health and welfare of the students. Given this context, why does it appear that there was a lack of recognition by those in charge that there needed to be an avenue for addressing the possible negative health effects on the residents of contaminated Superfund and Formerly Used Defense Site communities?

Consider again the River Valley graduates who suffered debilitating diseases and were left without a recourse for restitution because of government barriers that excluded a comprehensive approach that encompassed site cleanup and health benefit packages for those shown to be afflicted with cancer caused by contaminants on the school grounds, when the true latent periods are taken into consideration for cancers that may develop decades later. The lack of a comprehensive approach occurred for far too long, from 1997 through the present, a span of over twenty years. Will this barrier ever be eliminated to allow for a humanitarian solution to aid the afflicted communities affected by Superfund and Formerly Used Defense Sites? As it stands now, the only solution is a remedy through the court system, which has deterred many potential plaintiffs from challenging the government bureaucracy. Even if pursued, the litigious road is an expensive and time-consuming one that often leaves plaintiffs with large debts and no solutions. When will there be a consensus that real solutions include a recognition that cancer cases may be connected to contaminated sites and that restitution should include an examination of true latent periods that become a part of a comprehensive environmental site inspection and health benefit approach? You be the judge!

OBSERVATIONS

Ultimately, it appears that there was, tragically, a predetermination by the government agencies that they were not going to connect the debilitating diseases suffered by the River Valley Local School District graduates to the district's contaminated grounds. This outcome became clear when Crawford, a U.S. Army Corps of Engineers representative, stated in Paul Jayco's whistleblower hearing: "To be frank, our job is not to find the cause of the problem—by law our job is to look for any contamination at the Formerly Used Defense Site and to remediate it. Whether it causes cancer or not, we will clean it."[200]

This is a tragic story, an unfortunate legacy of how, by some accounts, the environmental data from the River Valley Local School District's environment site inspections remained uncollected, ignored or misinterpreted. We are thus left with the open question of what party or parties ought to have been held accountable for the contaminated school sites and made to pay restitution for the debilitating diseases incurred by the graduates of the River Valley Local School District. For sure, this situation has created many sleepless nights, even today, as graduates continue to fear when they might be the next victims of cancer that develops from exposure long ago to the cancer-causing agents identified in this book. Yet government agencies never have linked those cancer-causing agents to the graduates' incidence of cancer, which is the very reason why a health risk assessment reimbursement program is urgently needed to provide a fairness approach to reconcile future cancer cases that are attributed to contaminated Superfund and Formerly Used Defense Sites.

A pilot program for a hypothetical health risk assessment reimbursement proposal that includes an examination of true latent periods may be a start toward eliminating tort challenges and providing an avenue of compensation for individuals who can prove that their diseases were attributable to the contaminants found at Superfund and Formerly Used Defense sites. Such a program is akin to one instituted recently for the plaintiff in the Monsanto case.

It is hoped that this work provides those communities affected by Superfund and Formerly Used Defense Sites with a format that establishes a necessary knowledge base and understanding of the kinds of information that can be used to ask the right questions of government agencies and, at times, to challenge their assumptions to gain more clarity. Such an approach would follow these steps:

Examine Historical Evidence → Expand Community Action and Communication → Define the Results → Explore Legal Options

In addition to the River Valley Local School District history, there have been hundreds of thousands of people who have been negatively affected by Superfund and Formerly Used Defense Sites across the country, and those individuals have undoubtedly been dealt an unfair hand.

It is evident that there needs to be a change in how we deal with the health outcomes of contaminated communities. It appears that the only way this can happen is for affected communities to mobilize and to follow an action plan that pushes for adequate healthcare to counter those diseases that are likely to have been caused by the contaminated sites. Another part of this discussion centers on developing a health risk reimbursement program such as the one proposed in this book that could be established for both governmental and industrial contaminated sites. We are clearly at a tipping point, when one considers the Camp Lejeune settlement and Monsanto litigation, and the time is right for moving forward with an action plan that considers the health outcomes of those suffering from diseases related to contaminated sites.

It is perplexing that the outcome of the environmental site investigation reports was the conclusion that even though the leukemia rates of RVHS graduates was higher than the national average, government agencies continued to place the blame on higher rates of tobacco use. In a sense, the conclusions of the environmental site investigation reports of the River Valley Local School District come as no surprise, given that in the 1997 national survey of state health departments regarding cancer clusters, Ohio was the only state that did not participate in the survey. Recall the comment made by a U.S. Army Corps Engineers representative in the Paul Jayco hearing: "To be frank, our job is not to find the cause of the problem—by law our job is to look for any contaminants at this Formerly Used Defense Sites and to remediate it. Whether it causes cancer or not, we will clean it."[201] This statement foreshadowed the determination of government agencies to refrain from linking the high rates of leukemia and cancer of RVHS graduates to the contaminated school grounds.

In retrospect, was the science behind the studies of the contaminated school district sites flawed, as evidenced in chapter 4, which illustrates the misrepresentation of facts that showed the pathway through vinyl chloride was chloroethane and not chloroethene? This argument here matches the Agency for Toxic Substances and Disease Registry for vinyl chloride summary, which states: "Vinyl chloride is also known as chloroethene, chloroethylene and ethylene monochloride."[202] There is no mention of chloroethane being connected to vinyl chloride, and the chemical formulas are different, which leads one to believe that there was a misleading interpretation on the part of the government agencies. The bottom line, where is the justice? You be the judge!

Appendix A

RIVER VALLEY LOCAL SCHOOL DISTRICT SEQUENCE OF EVENTS

1942–1961—Marion Engineer Depot is the largest depot of its kind in the United States during World War II.

1944—Boxes of metals with protective wrapping that were designated for the Manhattan Project are stored at the Marion Engineer Depot.

1945—The Marion Engineer Depot stores fifty-eight thousand radium markers.

1947—A radiation survey indicates that twenty-three of twenty-six storage locations for metascopes—night vision equipment that contained radium—exceeded recommended radiation tolerance levels.

1955–1956—Radioactive materials from the metascopes are dumped somewhere. Could it have been on the Marion Engineer Depot grounds?

1957—A photo taken this year offers new clues: "River Valley schools could be resting on an old dump," the *Marion Star*, March 4, 1999.

December 1961—The River Valley Local School District purchases 59.10 acres of the former Marion Engineer Depot grounds for $7,350.

June 1964—The River Valley Local School District purchases 18.627 acres of the former Marion Engineer Depot grounds for $15,000.

June 1997—The Ohio Department of Health investigates a report of high cancer rates among River Valley High School graduates and students.

June 1997—The Concerned River Valley Families coalition is formed.

September 1997—A dime-sized disc coated in radium in found in the front yard of River Valley High School.

October 1997—A *Columbus Dispatch* headline reads "Leukemia News Stuns Marion—Death Rates up 122% in 30 Years, State Study Finds."

November 1997—Soil samples on the playing fields reveal high levels of polyaromatic hydrocarbons that included benzopyrene, a chemical linked to cancer. The finding is reaffirmed by former employees that the location was a former dumping site that included TCE fluids used to clean engine parts.

1998—A radiological survey finds there are no further areas of concern.

February 1998—High levels of arsenic are found in trenches near the high school; five athletic fields are roped off after finding more contamination.

August 1998—Restoration Advisory Board is formed to hold monthly meetings relative to the cleanup of the former Marion Engineer Depot and Scioto Ordnance Plant.

October 1998—Radiation is ruled as not a problem on the school district grounds. Bacteria from human and animal waste is found at the location of the former POW camp, where it was thought that this may have been the site of the POW latrine. The drinking fountains near the concession stands are found to have elevated levels of lead.

January 1999—U.S. Army Corps of Engineers confirms that the practice field areas were used as a waste dump.

February 1999—Ohio EPA and Ohio Department of Health declare there is no health risk on the school grounds. U.S. Army Corps of Engineers does not disclose that cancer-causing chemicals were detected during a dig on the athletic fields. Toxic barrels with leaking TCE and PCBs are found on the U.S. Army Reserve property next to the River Valley Local School District property line.

March 1999—Found aerial photos from 1957 indicate that the middle school may have been built on a toxic waste dump.

April 1999—A chain-link fence replaces the five-acre roped-off area of the athletic fields after more extensive pollution is found.

July 1999—Amended House Substitute Bill 282 creates line item 200-580: River Valley School Environmental Issues, which is used to assist with the costs arising from environmental assessment and cleanup of the potential environmental contamination of school facilities of the River Valley Local School District.

August 1999—Consultant Jed Ball claims that his radiation survey was doctored to show that the results would seem normal.

July 2000—Amended Substitute Senate Bill 272 modifies the Ohio School Facilities program to create the "Extreme Environmental Contamination of Classroom Facilities" section, which is targeted to assist the River Valley Local School District in moving up the priority list because of *extreme environmental contamination* and allowance of any federal restitution payments toward the project.

September 2000—Dr. Bruce Molholt, environmental toxicologist, says the students should be relocated because of the chemical health risks from the contaminated school grounds, but his report is rejected because the agencies said he mixed cancer data with environmental data.

October 2000—Paul Jayco's hearing at which a U.S. Army Corps of Engineers representative said: "To be frank, our job is not to find the cause of the problem—by law our job is to look for any contamination at the Formerly Used Defense Site and to remediate it. Whether it causes the cancer or not, we will clean it."

November 2000—Passage of a bond issue to participate in the Ohio School Facilities Program.

February 2000—Operable Unit 1 Report indicates that there is a high level of contaminants on the ballfields and athletic practice areas that contain high- and low-level carcinogens.

December 2000—Memorandum of Understanding and a Cooperative Agreement is signed by the Department of Army, the State of Ohio and the River Valley Local School District to share the cost of relocating the camp.

September 2001—Students are issued water bottles at the start of the 2001 school orientation and instructed to fill the bottles at home and bring them back to school.

December 2001—The Ohio Department of Health cannot pinpoint one particular factor that was associated with the development of leukemia among graduates and concludes that there is no need for further study. Expanded Site Inspection Report identifies SVOCS and TAL metals above screening levels on the main campus grounds.

January 2002—Ohio EPA releases a report that the indoor and outdoor air monitoring are consistent with the levels of prior reports and that there are no health risks to the students.

September 2003—Students are relocated to the new middle school and high school.

March 5, 2005—Twenty-seven thousand tons of contaminated soils with cancer-causing agents are removed from ballfields and athletic practice fields at an approximate cost of $8 million.

September 2008—Cleanup of the U.S. Army Reserve training center property is completed at an approximate cost of $2 million.

September 2012—Corps completes fourteen-year environmental cleanup of the former Marion Engineer Depot at an approximate cost of $35 million.

May 31, 2017—The *Columbus Dispatch* prints illustrated photos and an interview with former students who had cancer and who indicate that the government was at fault for selling property to the school district. One former student who was diagnosed with Hodgkin's disease in the 1980s said, "I don't know if there's anything you can do for the years. I have endured in living the medical nightmare I've had."

December 2018—*World War II POW Camps in Ohio* features a chapter that mentions the contamination of the River Valley Local School District, which became the premise for pursuing *The School Poisoning Tragedy in Caledonia, Ohio.*

Appendix B

ORGANIZATION ISSUES MONEY/POLITICS NEWS INDEX

electric competition
pollution prevention
pesticides

Vigil for students at River Valley Schools

Governor's Mansion, Parkview and Maryland in Bexley
8:00 p.m., Monday, August 21

Directions in Columbus: Take I-71 to the Broad Street exit. Go east, away from downtown, on Broad Street, past the Franklin Conservatory in Bexley. Turn left on Parkview. The Mansion is on the right side at the corner of Parkview and Maryland.

When school resumes this month, River Valley Middle School and High School students in Marion, Ohio will be forced onto school grounds contaminated with deadly toxic chemicals, unless Governor Bob Taft does something immediately.

- These schools were built on a site where the U.S. Army Marion Engineer Depot, the largest facility of its kind in the nation, dumped, burned and buried solvents, paints, fuel oil, and chemical agents for nearly two decades.
- Seventy different chemicals have been found on or near the school campus, including arsenic, vinyl chloride, trichloroethylene, benzo(a) pyrene, chromium, lead, PCBs, and solvents. In 1999, the U.S. Army found an imminent threat to human health on Army Reserve property near the schools.
- The Ohio Department of Health has confirmed a much-higher-than-expected rate of leukemia among River Valley graduates. The rate for esophageal cancer is ten times greater. Between 1966 and 1995, death rates from leukemia increased by 122% in the city of Marion, and brain cancer death rates increased by 40% in Marion County.

If a tornado roared through Marion and destroyed the River Valley Middle and High Schools, the state government would show up within hours to find a safe place for the children to learn. The River Valley Schools are just as uninhabitable, and the state government has spent years dragging its feet.

Join us to let Governor Taft know the people of Ohio care about these students and want him to close the schools and reassign the students to other schools immediately.

For information call Noreen Warnock, Ohio Citizen Action, (614) 263-4111.

Vigil for students at River Valley Schools. *Courtesy of the* Ohio Citizen, *August 21, 2000.*

NOTES

Preface

1. U.S. Army Corps of Engineers [hereafter USACE], "FUDS Inventory per State," September 30, 2015.
2. U.S. Environmental Protection Agency [hereafter EPA], "NPL Site Totals by Stuns and Milestone," April 19, 2019.

Introduction

3. National Archives and Records Administration [hereafter NARA image], *Depot Loading Area*.
4. NARA image, *Japanese Submarine*.

Chapter 1

5. *Columbus Dispatch*, October 17, 1997.
6. *Marion Star*, February 8, 1999.
7. U.S. General Accounting Office [hereafter GAO], "Ohio FUDS Cleanup Program."
8. Pawlik, "Erie Army Depot & Marion Engineer Depot."
9. Ohio Facilities Construction Commission, *Fiscal Year 2003 OSFC Annual Report*; USACE (Louisville), "Corps Completes 14-Year Environmental Cleanup at MED."
10. Associated Press, "Deal Struck to Move Ohio Schools."

11. An Act, Amended Substitute House Bill 282.
12. Ohio Legislative Service Commission, Amended Substitute Senate Bill 272.
13. USACE, "FUDS Inventory per State."
14. *Columbus Dispatch*, May 31, 2017.

Chapter 2

15. NARA images, *Sniper Scope Attached to a Rifle*; *Sniper Scope Attached to a Submachine Gun*; *Handheld Sniper Scope*.
16. Office of Legacy Management, "Radiological Survey, April 17, 1947."
17. Ibid.
18. Mosher and Mosher, *Scioto Ordnance Plant*, 108.
19. USACE (Rock Island), Archives Search Report.
20. *Columbus Dispatch*, October 24, 1997.
21. Ohio Citizen Action [hereafter OCA], "Recommended Decision and Preliminary Order."
22. NARA image, *Geiger Counter Sweeps*; OCA, "Recommended Decision and Preliminary Order."
23. *Columbus Dispatch*, October 24, 1997.
24. Mosher and Mosher, *Scioto Ordnance Plant*, 106.
25. NARA image, *Los Alamos*.
26. Battle Detective, "Battle Relic #13."
27. *Columbus Dispatch*, October 30, 1997.
28. *Columbus Dispatch*, August 28, 1999.
29. *Family Circle*, "Town Divided."
30. EPA, "U.S. EPA Radiation Education Activities," 9.
31. Ibid., 34.
32. Ibid., 13.
33. *Family Circle*, "Town Divided."
34. Patton, telephone interview, December 6, 2018.

Chapter 3

35. *Columbus Dispatch*, May 31, 2017.
36. *Columbus Dispatch*, September 11, 1997.
37. *Cincinnati Enquirer*, February 4, 1998.
38. *Columbus Dispatch*, September 13, 2018.
39. *Columbus Dispatch*, April 9, 1998.
40. *Marion Star*, February 19, 1999.
41. *Marion Star*, October 23, 1998.

42. *Family Circle*, "Town Divided."
43. *Marion Star*, October 23, 1998.
44. *Columbus Dispatch*, September 13, 2018.
45. *Columbus Dispatch*, February 18, 1999.
46. *Marion Star*, February 21, 1999.
47. *Columbus Dispatch*, September 26, 2008.
48. *Columbus Dispatch*, January 27, 1999.
49. *Columbus Dispatch*, February 12, 1999.
50. *Marion Star*, March 4, 1999.
51. Mosher and Mosher, *Scioto Ordnance Plant*, 123.
52. Van Keuren, *World War II POW Camps*.
53. Ibid.
54. Lord, interview.
55. USACE (Louisville), "Scioto Ordnance Plant."
56. OCA, "Ohio AFL-CIO Chief."
57. *Columbus Dispatch*, September 26, 2008.

Chapter 4

58. USACE, *Formerly Used Defense Site, Marion Engineer Depot* [hereafter *RVSD Property Report*], 225.
59. Ibid., 245.
60. Ibid. 2–10.
61. Ibid., 245.
62. Ibid., Figure 2-2, 200.1e.
63. Ibid., 210.
64. Ibid., 76.
65. Agency for Toxic Substances and Disease Registry [hereafter ATSDR], "Vinyl Chloride."
66. ATSDR, "Chloroethane."
67. Ibid.
68. ATSDR, "Vinyl Chloride."
69. USACE, *River Valley School Property Operable Unit 1* [hereafter *RVS Report*].
70. Ibid., Figure 4.2.
71. Ibid., Figure 4.3.
72. Ibid., Figures 4.1–4.4.
73. *Columbus Dispatch*, May 31, 2017.
74. *Family Circle*, "Town Divided."
75. *Columbus Dispatch*, September 26, 2008.
76. National Toxicology Program, "Trichloroethylene."

77. Ibid.
78. ATSDR, "Toxic Substance Portal," 2019.
79. ATSDR, "Substances Priority List," 2017.
80. ATSDR, "Toxic Substance Portal," 2019; ATSDR, "Substances Priority List," 2017.
81. *Marion Star*, February 8, 1999.

Chapter 5

82. USACE, *Restoration Advisory Board Adjournment Report*; *Marion Star*, February 8, 1999.
83. USACE, *Restoration Advisory Board Adjournment Report.*
84. *Marion Star*, October 6, 1998.
85. Ohio EPA, Media Advisory, July 14, 2000.
86. *Columbus Dispatch*, December 10, 1998.
87. *Columbus Dispatch*, August 6, 2000.
88. *Marion Star*, March 23, 1999.
89. OCA, "Vigil for Students."
90. Ohio EPA, Media Advisory, June 23, 2000.

Chapter 6

91. Ohio EPA, interviews, April 12, 2000 memo.
92. USACE, *RVSD Property Report*, 6–18.
93. NARA image, *Bailey Bridge across the Rhine River.*
94. NARA image, *Bailey Bridge across the Nisqually River.*
95. NARA images, *105mm Howitzer Practice Range*; *105mm Shell in the Breach Preparatory for Firing.*
96. NARA image, *Radiation Film Badge.*
97. *Family Circle*, "Town Divided."
98. Keller, email correspondence.

Chapter 7

99. *Columbus Dispatch*, July 28, 1997.
100. *Columbus Dispatch*, September 1, 1997.
101. *Cosmopolitan*, January 1999.
102. *Columbus Dispatch*, August 27, 1997.
103. *Columbus Dispatch*, April 30, 1998.
104. *Marion Star*, February 21, 1999.

105. *Columbus Dispatch*, October 7, 1998.
106. *Columbus Dispatch*, May 31, 2017.
107. *Columbus Dispatch*, June 9, 1998.
108. Patton, interview.
109. NARA, *Other Inspection Reports-Perry*.
110. Patton, interview.

Chapter 8

111. U.S. Department of Veteran Affairs [hereafter DVA], "Camp Lejeune."
112. OCA, "Judge's Decision."
113. *Marion Star*, February 2, 1999.
114. *Columbus Dispatch*, July 25, 2001.
115. *Marion Star*, February 2, 1999.
116. *Marion Star*, February 15, 1999.
117. *Marion Star*, February 16, 1999.
118. Ibid.
119. *Speak Ohio* post, June 17, 2000, 8:11 p.m.
120. *Speak Ohio* post, June 17, 2000, 3:39 p.m.
121. *Columbus Dispatch*, February 19, 1999.
122. *Columbus Dispatch*, February 21, 1999.
123. *Marion Star*, March 22, 1999.
124. *Columbus Dispatch*, September 30, 2000.
125. *Columbus Dispatch*, August 6, 2000.
126. *Speak Ohio* post, June 21, 2000, 11:37 a.m.
127. *Speak Ohio* post, June 21, 2000, 11:47 a.m.
128. USACE, interoffice communique, June 23, 2001.
129. Restoration Advisory Board minutes, May 23, 2002.
130. Marion Online, "Questions Raised."
131. *Columbus Dispatch*, March 5, 2001.
132. Ibid.
133. OCA, "Judge Rules Ohio EPA Illegally Punished."
134. *Toledo (OH) Blade*, "Whistleblower Reinstated."
135. GAO, "Environmental Contamination."
136. GAO, "Defense Infrastructure."
137. GAO, "Defense Infrastructure: DOD Can Improve."
138. GAO, "Military Base Realignments and Closures."
139. Restoration Advisory Board minutes, September 26, 2001.

Chapter 9

140. *Marion Star*, August 26, 1997.
141. *Marion Star*, October 23, 1997.
142. Ibid.
143. *Columbus Dispatch*, October 12, 1998.
144. Schettler, "Reverberations of Militarism," 8.
145. Ibid. 14–15.
146. National Center for Biotechnology Information, "Evaluation of Latent and Presumptive Periods."
147. Van Keuren, *River Valley Middle School Photograph*.
148. *Marion Star*, October 23, 1997
149. *Toledo (OH) Blade*, "River Valley Cancer Report."

Chapter 10

150. *Columbus Dispatch*, April 1, 1999.
151. *Speak Ohio* post, June 19, 2000, 7:05 p.m.
152. OCA, "Judge's Decision."
153. Restoration Advisory Board minutes, September 26, 2001.
154. Ibid.
155. Ibid.
156. Ibid.
157. Ibid.
158. EPA, Safe Drinking Water Act.

Chapter 11

159. Dutzik and Baumann, *Health Tracking & Disease Clusters*.
160. Harr, *No Safe Place*, 8.
161. Ibid., 11–13.
162. Science Education Resource Center Carleton College, "Science in the Courtroom."
163. *Columbus Dispatch*, February 22, 1999.
164. Ibid.
165. Van Keuren, *World War II POW Camps*, 2018.
166. Dutzik and Baumann, *Health Tracking & Disease Clusters*.
167. *Columbus Dispatch*, October 17, 1997.
168. Wilcox, "Hazardous Air Pollutants," 2.
169. Ohio Department of Health, *Case Review of Leukemia*, 13, 23.
170. Molholt, *Case Review*.

171. Ibid.
172. Ibid.
173. Ibid.
174. Ibid.
175. Ibid.
176. Ibid.
177. *Columbus Dispatch*, October 16, 1997.
178. CDC, "Cancer Clusters."
179. Ibid.
180. *Columbus Dispatch*, October 17, 1997.
181. *Columbus Dispatch*, February 22, 1999.
182. Wilcox, "Hazardous Air Pollutants."
183. ATSDR, "Substances Priority List."
184. Ohio EPA, internal memo, n.d.

Chapter 12

185. Patterson, interview.
186. *Columbus Dispatch*, May 31, 2017.
187. USACE, Cost of purchasing the school site, March 22, 2001.
188. Latham, Schwartz and Appel, "Intersection of Tort and Environmental Law," 752.
189. *Columbus Dispatch*, October 17, 1997.
190. EPA, "Superfund Liability."
191. DVA, "Health Issues."
192. EPA, "Conducting a Human Health Risk Assessment: Step 1."
193. EPA, "Conducting a Human Health Risk Assessment: Step 3."

Chapter 13

194. DVA, "Camp Lejeune."
195. NARA image, *Tornado*; Van Keuren, *Baseball Field Photograph*, 2019.
196. *Columbus Dispatch*, "Toxic Troubles," October 4, 1998.
197. *BuzzFeed News*, "Man Said He Got Cancer," March 19, 2019.
198. *New York Times*, "Toxic 'Forever Chemicals' in Drinking Water," February 22, 2019.
199. McGuire, Letter to OEPA and ODH, October 2000.
200. OCA, "Recommended Decision and Preliminary Order," 115.
201. Ibid., 54.
202. ATSDR, "Toxic Substance Portal," 2017.

BIBLIOGRAPHY

Agency for Toxic Substances and Disease Registry. "ATSDR Substances Priority List." https://www.atsdr.cdc.gov/spl.

———. "Chloroethane." https://www.atsdr.cdc.gov/substances/toxsubstance.asp?toxid=161.

———. "Toxic Substance Portal." https://www.atsdr.cdc.gov/substances/toxsubstances.asp?toxid=33.

———. "Vinyl Chloride." https://www.atsdr.cdc.gov/substances/toxsubstances.asp?toxid=51.

An Act, Amended Substitute House Bill 282. archives.legislature.state.oh.us/BillText123/123_HB_282_ENR.html.

Associated Press. "Deal Struck to Move Ohio Schools Off Tainted Soil." *Los Angeles Times*. http://www.latimes.com/archives/la-xpm-2000-may-20-mn-32125-story.html. May 20, 2000.

Battle Detective. "Battle Relic #13. Luminous Marker Disk." www.battledetective.com.

BuzzFeed News. "A Man Said He Got Cancer After Spraying Monsanto's Weed Killer. Jury Agreed." https://www.buzzfeednews.com.

Center for Disease Control and Prevention. "Cancer Clusters." https://www.cdc.gov/nceh/clusters/default.htm.

Cerny, George. Interviewed by Scott Shane and Sam Yebaile. [Telephone]. Ohio EPA–NWDO, Special Investigations Unit. Marion County Public Library, Marion, Ohio, March 1, 2000.

Cincinnati Enquirer. "Cancer-Causing Compounds Found at the Marion Co. School." FOIA P1L021.0345, Department of the Army, U.S. Army Corps of Engineers, Louisville, Kentucky, February 4, 1998.

Columbus Dispatch. "Another Cancer Joins list of Marion County concerns." FOIA P1L021.0348, Department of the Army, U.S. Army Corps of Engineers, Louisville, Kentucky, April 30, 1998.

———. "Army to Work with Group Looking into Marion Contamination." http://ohiocitizen.org/campaigns/marion/army_to_work.htm, July 25, 2001.

———. "Cancer Probe Taints First Day of School." FO1A P1L021.0306, Department of the Army, U.S. Army Corps of Engineers, Louisville, Kentucky, August 27, 1997.

———. "Cancer Rate Among School's Grads Prompts Investigation." FOIA P1L021.0295, Department of the Army. U.S. Army Corps of Engineers, Louisville, Kentucky, July 28, 1997.

———. "Chemicals Near School Called 'Imminent Threat." FOIA P1L021.0213, Department of the Army, U.S. Army Corps of Engineers, Louisville, Kentucky, February 12, 1999.

———. "Concerns Raised by Old Photo." FOIA P1L021.0200, Department of the Army, U.S. Army Corps of Engineers, Louisville, Kentucky, January 27, 1999.

———. "Building's Radiation Normal, Experts Say." FOIA P1L021.0322, Department of the Army, U.S. Army Corps of Engineers, Louisville, Kentucky, October 24, 1997.

———. "Corps Accused of Spinning Information." FOIA P1L021.0196, Department of the Army, U.S. Army Corps of Engineers, Louisville, Kentucky, December 10, 1998.

———. "DeWine Hears of Family's Plight." FOIA P1L021.0350, Department of the Army, U.S. Army Corps of Engineers, Louisville, Kentucky, June 9, 1998.

———. "Environmental Groups Want Schools Evacuated." FOIA P1L021.0220, Department of the Army, U.S. Army Corps of Engineers, Louisville, Kentucky, February 17, 1999.

———. "Findings Come as Surprise in Marion." FIOA P1L021.0363, Department of the Army, U.S. Corps of Engineers, Louisville, Kentucky, April 4, 1998.

———. "Folks Sick of Pollution Vow to Fight for Cleanup." FOIA P1L021.0370, Department of the Army, U.S. Army Corps of Engineers, Louisville, Kentucky, October 7, 1998.

———. "Higher Level of Poisons Detected." FOIA P1L021.0268, Department of the Army, U.S. Corps of Engineers, Louisville, Kentucky, April 9, 1999.

———. "Land Near Toxic River Valley Campus Cleaned Up Enough for Army Training." September 26, 2008.

———. "Leukemia News Stuns Marion." FOIA P1L021.0312, Department of the Army, U.S. Army Corps of Engineers, Louisville, Kentucky, October 17, 1997.

———. "Leukemia Rates Up in Marion." FOIA P1L021.0310, Department of the Army, U.S. Army Corps of Engineers, Louisville, Kentucky, October 16, 1997.

———. "Living with Doubt." FOIA P1L021.0224, Department of the Army, U.S. Army Corps of Engineers, Louisville, Kentucky, February 21, 1999.

———. "Marion County Parents Want Answers About School." FOIA P1L021.0296, Department of the Army, U.S. Army Corps of Engineers, Louisville, Kentucky, August 1, 1997.

———. "Marion Students at Risk, Expert Says." August 6, 2000.

———. "Mom's Mission Born of Son's Death from Leukemia." FOIA P1L021.0221, Department of the Army, U.S. Corps of Engineers, Louisville, Kentucky, February 18, 1999.

———. "New EPA Director Takes Charge of Marion Probe." FOIA P1L021.0222, Department of the Army, U.S. Army Corps of Engineers, Louisville, Kentucky, February 19, 1999.

———. "Photo of Workers Digging for Hazardous Materials at RVHS." Purchased from the Dispatch Printing Company, Norcross, Georgia, May 31, 2017.

———. "Questions Linger Decades after River Valley Cleanup." http://www.dispatch.com. May 31, 2017.

———. "Radiation Readings at School 10 Times Higher Than First Reported." FOIA P1L021.0362, Department of the Army, U.S. Army Corps of Engineers, Louisville, Kentucky, September 30, 1997.

———. "Radiation Source to Be Removed from the School Grounds." FOIA P1L021.0307, Department of the Army, U.S. Army Corps of Engineers, Louisville, Kentucky, September 11, 1997.

———. "Radiation Tester Says Findings at Marion Site Were Doctored an Investigator Says Supervisors Returned Altered Reports Altered Reports for Him and Others to Re-sign." www.ohiocitizen.org. August 29, 1999.

———. "River Valley Board Member Transfer Daughter." FOIA P1L021.0308, Department of the Army, U.S. Corps of Engineers, Louisville, Kentucky, August 6, 2000.

———. "State Agencies Dispute River Valley Findings." www.ohiocitizen.org. September 30, 2000.

———. "State Still Fighting Whistle-Blower Law." www.ohiocitizen.org. March 5, 2001.

———. "Surveyor Insists Marion Radiation Study Is Inaccurate." www.ohiocitizen.org. January 29, 2000.

———. "Toxic Shock—Deceptive Public Servants Shirk Duties." FOIA P1L021.0236, Department of the Army, U.S. Army Corps of Engineers, Louisville, Kentucky, February 21, 1999.

———. "Toxic Troubles—What Toll Taken on Public Health?" FOIA P1L021.0278, Department of the Army, U.S. Army Corps of Engineers, Louisville, Kentucky, October 4, 1998.

Cosmopolitan. "I Came Back from Death to Be a Cancer Crusader." FOIA P1L021.0209, Department of the Army, U.S. Corp of Engineers, Louisville, Kentucky, January 1999.

Danger Hazardous Waste sign. Purchased on eBay, January 3, 2019.

Dutzik, Tony, and Jeremiah Baumann. *Health Tracking & Disease Clusters.* U.S. PIRG Education Fund, Washington, D.C., September 2002.

Everly, Dwayne. Interviewed by Scott Shane [Telephone]. Ohio EPA–NWDO, Special Investigations Unit, Columbus, Ohio. Marion County Public Library, Marion, Ohio, March 6, 2000.

Family Circle. "A Town Divided: How a Cancer Threat Is Tearing a Community Apart." http://www.ohiocitizen.org/campaigns/marion/family_circle.html. August 2001.

Ferguson, Robert. Interviewed by Scott Shane and Samir Yebaile [One-on-One]. Ohio EPA–NWDO, Special Investigations Unit. Marion County Public Library, Marion, Ohio. February 18, 2000.

Gast, Loren. Interviewed by Scott Shane. [Telephone]. Ohio EPA–NWDO, Special Investigations Unit. Marion County Public Library, Marion, Ohio, March 2, 2000.

Gibbons, Harold. Interviewed by Scott Shane. [Telephone]. Ohio EPA–NWDO, Special Investigations Unit. Marion County Public Library, Marion, Ohio, March 2, 2000.

Gorrell, Donald. Interviewed by Scott Shane and Samir Yebaile [One-on-One]. Ohio EPA–NWDO, Special Investigations Unit. Marion County Public Library, Marion, Ohio, February 1, 2000.

Harr, Jonathan. *No Safe Place—Toxic Waste, Leukemia, and Community Action.* Berkeley: University of California Press, 1990.

Hill, Ralph, Jr. Interviewed by Scott Shane and Samir Yebaile [One-on-One]. Ohio EPA–NWDO, Special Investigations Unit. Marion County Public Library, Marion, Ohio, February 1, 2000.

Johnson, John. Interviewed by Scott Shane [Telephone]. Ohio EPA–NWDO, Special Investigations Unit, March 29, 2000.

Keller, Gary. Email correspondence. Marion Engineer Depot. February 25, 2019.

Latham, Mark, V.E. Schwartz and C.E. Appel. "The Intersection of Tort and Environmental Law: Where the Twain Should Meet and Depart." *Fordham Law Review* 80, no. 2 (2011). https://ir.lawnet.fordham.edu/flr/vol80/iss2/12.

Livingston, Bill. Interviewed by Scott Shane and Samir Yebaile [One-on-One]. Ohio EPA–NWDO, Special Investigations Unit. Marion County Public Library, Marion, Ohio, January 20, 2000.

Lord, Norman. Interviewed by Scott Shane [Telephone]. Ohio EPA–NWDO, Special Investigations Unit, Columbus, Ohio. Marion County Public Library, Marion, Ohio, March 10, 2000.

Lowery, Bob. Interviewed by Scott Shane and Samir Yebaile [One-on-One]. Ohio EPA–NWDO, Special Investigators Unit, Columbus, Ohio. Marion County Public Library, Marion, Ohio, February 8, 2000.

Marion Online. "Questions Raised About RV Remediation Costs." Mariononline.com FOIA Department of the Army, U.S. Corps of Engineers, Louisville, Kentucky, June 7, 2000.

Marion Star. "Anger Mounts Over Lack of Action." FOIA P1L021.0214, Department of the Army, U.S. Army Corps of Engineers, Louisville, Kentucky, February 13, 1999.

———. "Army Reserve Property." FOIA P1L021.0227, Department of the Army, U.S. Army Corps of Engineers, Louisville, Kentucky, February 21, 1999.

———. "Better Safe Than Sorry." FOIA P1L021.0248, Department of the Army, U.S. Army Corps of Engineers, Louisville, Kentucky, March 23, 1999.

———. "Counting Cases." FOIA P1L021.0229. Department of Army, U.S. Army Corps of Engineers, Louisville, Kentucky, February 22, 1999.

———. "Fallout from Toxic Barrels Continues." FOIA P1L021.02215, Department of the Army, U.S. Army Corps of Engineers, Louisville, Kentucky, February 15, 1999.

———. "Free Cancer Prevention Forums Are Offered." FOIA P1L021.0353, Department of the Army, U.S. Corps of Engineers, Louisville, Kentucky, October 6, 1998.

———. "Health Board to Hire Full-Time Administrator." FOIA P1L021.0238, Department of the Army, U.S. Corps of Engineers, Louisville, Kentucky, February 18, 1999.

———. "Leukemia Update: Letter to Parents." FOIA P1L021.0216, Department of the Army, U.S. Army Corps of Engineers, Louisville, Kentucky, February 16, 1999.

———. "Marion Deserves Better from the Army Corps." FOIA P1L021.0217, Department of the Army, U.S. Army Corps of Engineers, Louisville, Kentucky, February 16, 1999.

———. "OEPA, Samples School Soil Today." FOIA P1L021.0218, Department of the Army, U.S. Army Corps of Engineers, Louisville, Kentucky, February 16, 1999.

———. "Photo from 1957 Offers New Clues." FOIA P1L021.0240, Department of the Army, U.S. Corps of Engineers, Louisville, Kentucky, March 4, 1999.

———. "Private Report: Remove Barrels Now." February 19, 1999. FOIA P1L021.0226, Department of the Army, U.S. Army Corps of Engineers, Louisville, Kentucky, February 19, 1999.

———. "River Valley Graduates and Families, We Need Your Help." FOIA P1L021.0239, Department of the Army, U.S. Army Corps of Engineers, Louisville, Kentucky, February 8, 1999.

———. "Self-Reported Cancer Case Numbers Rise." FOIA P1L021.0228, Department of the Army, U.S. Army Corps of Engineers, Louisville, Kentucky, February 22, 1999.

———. "U.S. Army Reserves Knew about Toxic Barrels 3 Years Ago." FOIA P1L021.0249, Department of the Army, U.S. Army Corps of Engineers, Louisville, Kentucky, March 21, 1999.

———. "Valley Grads Troubled by Agency Letter." FOIA P1L021.0212, Department of the Army, U.S. Army Corps of Engineers, Louisville, Kentucky, February 2, 1999.

———. "Valley Tests Show High Metal Levels." FOIA P1L021.0364, Department of the Army, U.S. Army Corps of Engineers, Louisville, Kentucky, October 23, 1998.

McGuire, James. Letter to the Ohio EPA and Ohio Department of Health. Technical Subcommittee of the River Valley Schools Restoration Advisory Board. Marion County Public Library, Marion, Ohio, October 2000.

McKnight, Melvin. Interviewed by Scott Shane and Samir Yebaile [One-on-One]. Ohio EPA–NWDO, Special Investigations Unit. Marion County Public Library, Marion, Ohio, 18, 2000.

Molholt, Bruce. *Case Review of Leukemia Among Residents of Marion County, Ohio, 1992–1999, and Graduates of River Valley High School, 1963-2000 (ODH)*. Marion, OH: Marion County Public Library, 2001.

———. Memorandum to J.R. Kolmer, P.E. September 17, 2001.

Mosher, Charles D., and Delpha R. Mosher. *The Scioto Ordnance Plant and the Marion Engineer Depot of Marion, Ohio: A Profile After Forty Year.* Wauseon, OH: Glanz Printing Company, 1987.

National Archives and Records Administration. *105mm Howitzer Practice Range.* Photo: No. 111-SC-251529. College Park, Maryland, March 6, 1944.

———. *105mm Shell in the Breach Preparatory for Firing.* Photo: No. 111-SC-251530. College Park, Maryland, 1944.

———. *Bailey Bridge across the Nisqually River.* Photo: No. SC 327911. College Park, Maryland, October 31, 1951.

———. *Bailey Bridge across the Rhine River.* Photo: No. SC 32911. College Park, Maryland, April 18, 1945.

———. *Depot Loading Area.* College Park, Maryland, July 27, 1943.

———. *Geiger Counter Sweeps.* Photo: No. 111-SC- 427254. College Park, Maryland, July 25, 1953.

———. *Handheld Sniper Scope.* Photo: No. SC 464440. College Park, Maryland, June 29, 1954.

———. *Japanese Submarine at the Marion Engineer Depot.* Photo: Nos. SC 339586 & SC 339587. College Park, Maryland, July 2, 1943.

———. *The Loading Section at the Marion Engineer Depot.* Photo: No. SC 339601. College Park, Maryland, July 27, 1943.

———. *Los Alamos, New Mexico Scientific Laboratories.* Photo: No. 306-PS-54-8361. College Park, Maryland, June 11, 1954.

———. *Other Inspection Reports-Perry.* RG 389, Entry 461. Box 2669 @Stack Location 290/34/28/5. College Park, Maryland, 1944-1945.

———. *Radiation Film Badge.* Photo: No. 22118242. College Park, Maryland, June 21, 1948.

———. *Sniper Scope Attached to a Rifle.* Photo: No. SC-239417, College Park, Maryland, January 1, 1946.

———. *Sniper Scope Attached to a Submachine Gun.* Photo: No. SC-261501, November 12, 1946.

———. *Tornado–Greensburg, Kansas Burn Pit.* Photo: RG 311. College Park, Maryland, May 19, 2007.

National Center for Biotechnology Information. "Evaluation of Latent and Presumptive Periods—Veterans and Agent Orange." https://www.ncbi.nlm.nih.gov.

National Toxicology Program. "Trichloroethylene." https://ntp.niehs.nih.gov.

New York Times. "Toxic 'Forever Chemicals' in Drinking Water Leave Military Families Reeling." February 22, 2019.

Office of Legacy Management. "Radiological Survey, April 17, 1947." https://www.lm.doe.gov.

Ohio Citizen Action. "Judge Rules Ohio EPA Illegally Punished Investigator." http://www.ohiocitizen.org.

———. "Judge's Decision on Jayco/River Valley Case—Part Two." http://www.ohiocitizen.org.

———. "Ohio AFL-CIO Chief: School District Should 'Move the River Valley Students Immediately'." http://www.ohiocitizen.org.

———. "Recommended Decision and Preliminary Order. Paul Jayco v. Ohio EPA—Part Two." http://www.ohiocitizen.org.

———. "Vigil for Students at River Valley Schools." FOIA P1L021.0376, Department of the Army, U.S. Army Corps of Engineers, Louisville, Kentucky, August 21, 2000.

Ohio Department of Health. *Case Review of Leukemia Among Residents of Marion County, Ohio, 1992–1999 and Graduated of River Valley High School, 1963–2000.* Final Report, Marion County Public Library, Marion, Ohio, July 23, 2001.

Ohio EPA. Internal Memo (n.d.). FOIA P1L021.0394 Department of the Army, U.S. Corps of Engineers, Louisville, Kentucky.

———. Interviews Regarding Waste at RVHS and MED. Marion Public Library, Marion, Ohio, April 12, 2000.

———. Media Advisory—River Valley Schools/Marion Investigation. FOIA P1L021.0387 Department of the Army, U.S. Army Corps of Engineers, Louisville, Kentucky, June 23, 2000.

———. Media Advisory—River Valley Schools/Marion Investigation. FOIA P1L021.0293 Department of the Army, U.S. Army Corps of Engineers, Louisville, Kentucky, July 14, 2000.

Ohio Facilities Construction Commission. *Fiscal Year 2003 OSFC Annual Report.* https://ofcc.ohio.gov/Resources/Publications.

———. *Fiscal Year 2017 OFCC Annual Report.* https://ofcc.ohio.gov/Resources/Publications.

Ohio Legislative Service Commission. Amended Substitute Senate Bill 272. www.lsc.Ohio.gov/analyses/s0272-rh.pdf.

Patterson, Don. Interviewed by Scott Shane and Samir Yebaile [One-on-One]. Ohio EPA–NWDO, Special Investigations Unit. Marion County Public Library, Marion, Ohio, February 17, 2000.

Patton, Patrick. Interviewed by James Van Keuren [Telephone]. December 6, 2018.

Pawlik, Eugene. "Erie Army Depot & Marion Engineer Depot." Email. September 16, 2018.

Restoration Advisory Board. Meeting Minutes. Marion County Public Library, Marion, Ohio, September 26, 2001.

———. Meeting Minutes–Marion Engineer Depot and Scioto Ordnance Plant. FOIA P1L021.0428, Department of the Army, U.S. Army Corps of Engineers, Louisville, Kentucky, May 23, 2002.

Schettler, Ted H. "Reverberations of Militarism: Toxic Contamination, the Environment, and Health." International Physicians for the Prevention of Nuclear War. https://www.ippnw.org/pdf/mgs/2-1-schettler.pdf

Science Education Resource Center Carleton College. "Science in the Courtroom: The Woburn Toxic Trial." https://serc.carleton.edu/index.html.

Speak Ohio. "Another School & City View of River Valley." FOIA P1L021.0388, Department of the Army, U.S. Army Corps of Engineers, Louisville, Kentucky, June 17, 2000, at 8:11 p.m.

———. "Another School & City View of River Valley." FOIA P1L021.0388, Department of the Army, U.S. Army Corps of Engineers, Louisville, Kentucky, June 17, 2000 at 3.39 p.m.

———. "Marion Industrial Depot's New Water Line." FOIA P1L021.0388, Department of the Army, U.S. Army Corps of Engineers, Louisville, Kentucky, June 19, 2000 at 7:05 p.m.

———. "Watching and Wondering." FOIA P1L021.0388, Department of the Army, U.S. Army Corps of Engineers, Louisville, Kentucky, June 21, 2000 at 11:37 a.m.

———. Watching and Wondering." FOIA P1L021.0388, Department of the Army, U.S. Army Corps of Engineers, Louisville, Kentucky, June 21, 2000 at 11:47 a.m.

Timmer, Richard. Interviewed by Scott Shane [Telephone]. Ohio EPA–NWDO, Special Investigations Unit. Marion County Public Library, Marion, Ohio, March 6, 2000.

Tincher, John. Interviewed by Samir Yebaile [Telephone]. Ohio EPA–NWDO, Special Investigations Unit. Marion County Public Library, Marion, Ohio, March 9, 2000.

Toledo (OH) Blade. "River Valley Cancer Report Attached." FOIA P1L021.0366, Department of the Army, U.S. Army Corps of Engineers, Louisville, Kentucky, July 28, 2001.

———. "Whistleblower Reinstated." February 24, 2001.

U.S. Army Corps of Engineers. *Formerly Used Defense Site, Marion Engineer Depot, Marion, Ohio–River Valley School Property Expanded Site Inspection Report*. FOIA 13-0090, Department of the Army, U.S. Army Corps of Engineers, Louisville, Kentucky, December 2001.

———. "FUDS Inventory per State." https://www.usace.army.mil.

———. Interoffice communique. FOIA P1L021.0418, Department of the Army, U.S. Army Corps of Engineers, Louisville, Kentucky, June 29, 2001.

———. *Restoration Advisory Board Adjournment Report Former Marion Engineer Depot and Scioto Ordnance Plant Marion, Ohio*, Louisville, Kentucky, July 2010.

———. *Restoration Advisory Board Notice*. P1L021.0423, Department of the Army, U.S. Army Corps of Engineers, Louisville, Kentucky, May 23, 2002.

———. *River Valley School Property Operable Unit 1— Former Disposal Area Draft Remedial Investigation*. Vol. 1. Marion, OH: Marion Public Library, 2000.

U.S. Army Corps of Engineers (Louisville District). "Corps Completes 14-Year Environmental Cleanup of Marion Engineer Depot." https://www.lrl.usace.army.mil.

———. Cost of Purchasing the School Site from the Marion Engineer Depot. FOIA P1L021.0395, Department of the Army, U.S. Army Corps of Engineers, Louisville, Kentucky, May 22, 2001.

———. *Photo of Environmental Cleanup of the River Valley Local School District*. https://www.lrl.usace.mil/Missions/Environmental/Marion-Engineer-Depot, March 5, 2005.

———. "Scioto Ordnance Plant." lrl.usace.army.mil/Mission/Environmental/Scioto-Ordnance-Plant.

U.S. Army Corps of Engineers (Rock Island District), Archives Search Report for the former Marion Engineer Depot (F-8). Marion County Public Library, Marion, Ohio, July 1998.

U.S. Department of Veterans Affairs. "Camp Lejeune: Past Water Contamination." https://www.publichealth.va.gov.

———. "Health Issues Related to Camp Lejeune Water Contamination." https://www.va.gov.

U.S. Environmental Protection Agency. "Conducting a Human Health Risk Assessment: Step 1: The 4 Step Risk Assessment Process." https://www.epa.gov.

———. "Conducting a Human Health Risk Assessment: Step 3: Exposure Assessment." https://www.epa.gov.

———. "Drinking Water Contaminants—Standards and Regulations." https://www.epa.gov.

———. "NPL Site Totals by Status and Milestone." https://www.epa.gov.

———. Safe Drinking Water Act (SDWA). https://www.epa.gov/sdwa.

———. "Superfund Liability." https://www.epa.gov.

———. "U.S. EPA Radiation Education Activities: Radiation Exposure." https://www3.epa.gov/radtown/docs/radiation-exposure-activities.pdf.

U.S. General Accounting Office. "Environmental Contamination: DOD Has Taken Steps to Improve Cleanup Coordination at Former Defense Sites but Clearer Guidance Is Needed to Ensure Consistency." https://www.gao.gov/products/GAO-03-146. March 28, 2003.

———. "Defense Infrastructure: DOD Can Improve Its Response to Environmental Exposures on Military Installations." https://www.gao.gov/products/GAO-12-412. May 1, 2012.

———. "Defense Infrastructure: Interagency Agreements and Improved Project Management Needed to Achieve Cleanup Progress at Key Defense Installations." https://www.gao.gov/products/GAO-10-348. July 15, 2010.

———. "Military Base Realignments and Closures: DOD Has Improved Environmental Cleanup Reporting but Should Obtain and Share More Information." https://www.gao.gov/products/GAO-17-151. January 19, 2017.

———. "Ohio FUDS Cleanup Program." https: www.gov/gao.gov/gao-01-1012sp/OH.html. July 31, 2001.

———. "Superfund: Interagency Agreements and Improved Project Management Needed to Achieve Cleanup Progress at Key Defense Installations." https://www.gao.gov. July 15, 2010.

Van Keuren, James. *Baseball Field Photograph.* Put-in-Bay, Ohio, 2019.

———. *River Valley High School Photograph.* Caledonia, Ohio, 2019.

———. *River Valley Middle School Photograph.* Caledonia, Ohio, 2019.

———. *World War II POW Camps in Ohio*. Charleston, SC: The History Press, 2018.

Wilcox, Patricia P. "Hazardous Air Pollutants and Deaths Due to Lymphatic and Hematopoietic Disorders in Ohio, 1988–1997." Master's thesis, The Ohio State University. FOIA P1L021.0430, Department of the Army, U.S. Army Corps of Engineers, Louisville, Kentucky, 2003.

INDEX

O

P

R

S

T

U

V

W

ABOUT THE AUTHOR

Dr. James Van Keuren, EdD, is a retired professor of educational administration and dean of the Dwight Schar College of Education at Ashland University in Ashland, Ohio, where he published numerous juried journal articles, including a book titled *Web-Based Instruction: A Practical Guide for Online Courses.* Prior to Ashland University, Dr. Van Keuren was a teacher, high school principal, school superintendent, assistant and interim state superintendent of public instruction in Ohio. In *A Tribute to the 109th Evacuation Hospital (SM)*, he traces his father-in-law's service time during World War II with a second edition reflecting the integral role that nurses played in the unit. His most recent book, *World War II POW Camps in Ohio,* examines the Germans and Italians who were housed in Ohio during World War II. Dr. Van Keuren has been married to his wife, Pat, for fifty years, and they have daughters Michelle Scott and Christie Murdoch and sons-in-law Drew Scott and Dale Murdoch, in addition to grandchildren Hailey Scott, Weston Murdoch and Cooper Murdoch.